Lecture Notes in Computer Science

Commenced Publication in 1973
Founding and Former Series Editors:
Gerhard Goos, Juris Hartmanis, and Jan van Leeuwen

Guangyan Huang Xiaohui Liu Jing He
Frank Klawonn Guiqing Yao (Eds.)

Health Information Science

Second International Conference, HIS 2013
London, UK, March 25-27, 2013
Proceedings

Volume Editors

Guangyan Huang
Victoria University, Melbourne, VIC, Australia
E-mail: guangyan.huang@vu.edu.au

Xiaohui Liu
Brunel University, London, UK
E-mail: xiaohui.liu@brunel.ac.uk

Jing He
Victoria University, Melbourne, VIC, Australia
E-mail: jing.he@vu.edu.au

Frank Klawonn
Ostfalia University of Applied Sciences
Wolfenbüttel, Germany
E-mail: f.klawonn@ostfalia.de

Guiqing Yao
University of Southampton, UK
E-mail: g.yao@soton.ac.uk

ISSN 0302-9743   e-ISSN 1611-3349
ISBN 978-3-642-37898-0   e-ISBN 978-3-642-37899-7
DOI 10.1007/978-3-642-37899-7
Springer Heidelberg Dordrecht London New York

Library of Congress Control Number: 2013935527

CR Subject Classification (1998): H.4, H.3, I.2, H.2.8, J.1, I.5, K.6.5

LNCS Sublibrary: SL 3 – Information Systems and Application, incl. Internet/Web
and HCI

Typesetting: Camera-ready by author, data conversion by Scientific Publishing Services, Chennai, India

Printed on acid-free paper

Springer is part of Springer Science+Business Media (www.springer.com)

Preface

The International Conference Series on Health Information Science (HIS) provides a forum for disseminating and exchanging multidisciplinary research results in computer science/information technology and health science and services. It covers all aspects of health information sciences and systems that support health information management and health service delivery.

Following the successful first HIS conference held in Beijing in April 2012, the Second International Conference on Health Information Science (HIS 2013) was held in London, UK, during March 25-27, 2013. The scope of the conference includes (1) medical/health/biomedicine information resources, such as patient medical records, devices and equipments, software and tools to capture, store, retrieve, process, analyze, and optimize the use of information in the health domain, (2) data management, data mining, and knowledge discovery, all of which play a key role in decision making, management of public health, examination of standards, privacy and security issues, and (3) development of new architectures and applications for health information systems.

The HIS 2013 program had three keynote speakers; Joost N. Kok from the Leiden Institute of Advanced Computer Science, Leiden University, The Netherlands; Dame Julie Moore from University Hospitals Birmingham, UK; Jinhai Shi from Tianjin International Joint Academy of Biotechnology and Medicine, China. Based on the Program Committee's informative reviews, a total of 27 submissions were accepted for presentation at the conference, including 20 full papers, three short papers, three demo papers, and one poster. The authors are from 18 countries, and some will be invited to submit the extended versions of their papers to a special issue of the *Health Information Science and System Journal*, published by BioMed Central (Springer).

We highly appreciate the reviews provided by the Program Committee members. The conference would not have been successful without the support of the Organizing Committee members, invited speakers, authors, and conference attendees.

Our special thanks go to the host organizations: Brunel University (UK), Victoria University (Australia), and the UCAS-VU Joint Lab for Social Computing and E-Health Research, the University of Chinese Academy of Science (China), and the local arrangements team and volunteers.

We would like to sincerely thank our financial supporters and sponsors: the Multidisciplinary Assessment of Technology Centre for Healthcare (MATCH) and UCAS-VU Joint Lab for Social Computing and E-Health Research.

Finally we acknowledge all those who contributed to the success of HIS 2013 but whose names cannot be listed here.

March 2013

Terry Young
Michelle Towstoless
Yanchun Zhang
Xiaohui Liu
Jing He
Frank Klawonn
Guiqing Yao

Organization

General Co-chairs

Terry Young — Brunel University, UK
Michelle Towstoless — Victoria University, Australia
Yanchun Zhang — Victoria University, Australia and University of Chinese Academy of Science, China

Program Co-chairs

Xiaohui Liu — Brunel University, UK
Jing He — Victoria University, Australia
Frank Klawonn — Ostfalia University of Applied Sciences, Germany
Guiqing Yao — University of Southampton, UK

Workshop Co-chairs

Brian Meenan — University of Ulster, UK
Steve Morgan — University of Nottingham, UK

Industrial Chair

Loy Lobo — BT Global Services

Demo Chair

Chaoyi Pang — Australia e-Health research Centre, Australia

Financial Chair

Louise Dudley — Brunel University, UK

Publication Chair

Guangyan Huang — Victoria University, Australia

Publicity Co-chairs

Qixin Wang — The Hong Kong Polytechnic University, China
Fernando J. Martin-Sanchez — Melbourne University, Australia

Event and Venue Manager

Elizabeth Deadman   Brunel University, UK

Geographic Area Chairs

Leonard Goldschmidt   Stanford University Medical School, USA
William Wei Song   Dalarna University, Sweden
Michael Steyn   Royal Brisbane & Women's Hospital, Australia
Jie Liu   Bejing Jiao Tong University, China

Program Committee

Michael Berthold   University of Konstanz, Germany
Ilvio Bruder   University of Rostock, Germany
Klemens Böhm   Karlsruhe Institute of Technology, Germany
Carlo Combi   University of Verona, Italy
Carole Cummins   University of Birmingham, UK
Elizabeth Deadman   Brunel University, UK
Patrik Eklund   Umea University, Sweden
Ling Feng   Tsinghua University, China
Alex Gammerman   Royal Holloway, University of London, UK
Matjaz Gams   Jožef Stefan Institute, Slovenia
Lorenz Grigull   Hannover Medical School, Germany
Yi-Ke Guo   Imperial College London, UK
Reinhold Haux   Technical University of Braunschweig,
                             Germany
Jing He   Victoria University, Australia
Georg Hoffmann   Institut für Molekulare Onkologie, Germany
Long Jiao   Imperial College London, UK
Frank Klawonn   Ostfalia University of Applied Sciences,
                             Germany
Petra Knaup-Gregori   University of Heidelberg, Germany
Girard Krause   Robert Koch Institute, Germany
Hui Li   University of Birmingham, UK
Yongmin Li   Brunel University, UK
Jinling Liang   Southeast University, China
Xiaohui Liu   Brunel University, UK
Yurong Liu   Yangzhou University, China
Zhiyuan Luo   Royal Holloway, University of London, UK
Nigel Martin   University of London, UK
Sally McClean   Ulster University, UK

Fionn Murtagh	University of London, UK
Frank Pessler	Helmholtz Centre for Infection Research, Germany
Jose M. Peña	Technical University of Madrid, Spain
Ren Ran	Dalian Medical University, China
Thomas Rattei	University of Vienna, Austria
Lucia Sacchi	University of Pavia, Italy
Paola Sebastiani	Boston University , USA
Jinhai Shi	Tianjin International Joint Academy of Biotechnology and Medicine, China
Weiqing Sun	University of Toledo, USA
Veronica Vinciotti	Brunel University, UK
Qixin Wang	The Hong Kong Polytechnic University, Hong Kong, China
Zidong Wang	Brunel University, UK
Liang Xiao	Nanjing University of Science and Technology, China
Juanying Xie	Shaanxi Normal University, China
Liangdi Xie	Fujian Medical University, China
Shengxiang Yang	De Montfort University, UK
Terry Young	Brunel University, UK
Wenhua Zeng	Xiamen University, China
Huiru Zhang	University of Ulster, UK
Jenny Zhang	RMIT University, Australia
Yanchun Zhang	Victoria University, Australia
Zili Zhang	Deakin University, Australia
Zijun Zhou	Peking University, China
Shihua Zhu	University of Birmingham, UK

Steering Committee

Chairs

| Yanchun Zhang | Victoria University, Australia and University of Chinese Academy of Science, China |

Members

Terry Young	Brunel University, UK
Michael Styen	Royal Brisbane and Women's hospital, Australia
Jie Liu	Beijing Jiaotong University, China
Leonard Goldschmidt	Stanford University Medical School, USA
Jing He	Victoria University, Australia

Table of Contents

Modeling and Validating the Clinical Information Systems Policy Using Alloy

Ramzi A. Haraty and Mirna Naous

Department of Computer Science and Mathematics
Lebanese American University
Beirut, Lebanon
rharaty@lau.edu.lb

Abstract. Information systems security defines three properties of information: confidentiality, integrity, and availability. These characteristics remain major concerns throughout the commercial and military industry. In this work, we focus on the integrity aspect of commercial security applications by exploring the nature and scope of the famous integrity policy - the Clinical Information Systems Policy. We model it and check its consistency using the Alloy Analyzer.

Keywords: Clinical Information Systems Model, Consistency, and Integrity.

1   Introduction

The goal of information systems is to control or manage the access of subjects (users, processes) to objects (data, programs). This control is governed by a set of rules and objectives called a security policy. Data integrity is defined as "the quality, correctness, authenticity, and accuracy of information stored within an information system" [1]. Systems integrity is the successful and correct operation of information resources. Integrity models are used to describe what needs to be done to enforce the information integrity policies. There are three goals of integrity:

- Prevent unauthorized modifications,
- Maintain internal and external consistency, and
- Prevent authorized but improper modifications.

Before developing a system, one needs to describe formally its components and the relationships between them by building a model. The model needs to be analyzed and checked to figure out possible bugs and problems. Thus, formalizing integrity security models helps designers to build a consistent system that meets its requirements and respects the three goals of integrity. This objective can be achieved through the Alloy language and its analyzer.

Alloy is a structural modeling language for software design. It is based on first order logic that makes use of variables, quantifiers and predicates (Boolean functions) [2]. Alloy, developed at MIT, is mainly used to analyze object models. It translates constraints to Boolean formulas (predicates) and then validates them using the Alloy

G. Huang et al. (Eds.): HIS 2013, LNCS 7798, pp. 1–17, 2013.

Analyzer by checking code for conformance to a specification [3]. Alloy is used in modeling policies, security models and applications, including name servers, network configuration protocols, access control, telephony, scheduling, document structuring, and cryptography. Alloy's approach demonstrates that it is possible to establish a framework for formally representing a program implementation and for formalizing the security rules defined by a security policy, enabling the verification of that program representation for adherence to the security policy.

There are several policies applied by systems for achieving and maintaining information integrity. In this paper, we focus on the Clinical Information Systems Security Policy [4] and to show how it can be checked for consistency or inconsistency using the Alloy language and the Alloy Analyze.

The remainder of this paper is organized as follows: Section 2 provides the literature review. Section 3 discusses the Clinical Information System Security Model, and section 4 concludes the paper.

2   Literature Review

Hassan and Logrippo [5] proposed a method to detect inconsistencies of multiple security policies mixed together in one system and to report the inconsistencies at the time when the secrecy system is designed. The method starts by formalizing the models and their security policies. The mixed model is checked for inconsistencies before real implementation. Inconsistency in a mixed model is due to the fact that the used models are incompatible and cannot be mixed. The authors demonstrated their method by mixing the Bell-LaPadula model [6] with the Role Based Access Control (RBAC) model [7] in addition to the Separation of Concerns model [8]. Two modes are used for combination of models: mixed and hybrid. The system that adopts mixed-mode secrecy implements policies following any parent model. Mixed models combine the parent model's policies and their properties. On the other hand, hybrid models inherit properties from parent models or include other properties not available in the parent. In a mixed secrecy model there is always inconsistency. The authors addressed two types of inconsistencies: model, and system. Model inconsistency is the logical conflict between properties and meta policies. System inconsistency is the conflict between user policies or between user policies and meta policies.

Zao et al. [9] developed the RBAC schema debugger. The debugger uses a constraint analyzer built into the lightweight modeling system to search for inconsistencies between the mappings among users, roles, objects, permissions and the constraints in a RBAC schema. The debugger was demonstrated in specifying roles and permissions according and verifying consistencies between user roles and role permissions and verifying the algebraic properties of the RBAC schema.

Hassan et al. [10] presented a mechanism to validate access control policy. The authors were mainly interested in higher level languages where access control rules can be specified in terms that are directly related to the roles and purposes of users. They discussed a paradigm more general than RBAC in the sense that the RBAC can be expressed in it.

Shaffer [11] described a security Domain Model (DM) for conducting static analysis of programs to identify illicit information flows, such as control dependency flaws and covert channel vulnerabilities. The model includes a formal definition for trusted subjects, which are granted privileges to perform system operations that require mandatory access control policy mechanisms imposed on normal subjects, but are trusted not to degrade system security. The DM defines the concepts of program state, information flow and security policy rules, and specifies the behavior of a target program.

Misic and Misic [12] addressed the networking and security architecture of healthcare information system. This system includes patient sensor networks, wireless local area networks belonging to organizational units at different levels of hierarchy, and the central medical database that holds the results of patient examinations and other relevant medical records. In order to protect the integrity and privacy of medical data, they proposed feasible enforcement mechanisms over the wireless hop.

Haraty and Boss [13] showed how secrecy policies can be checked for consistency and inconsistency by modeling the Chinese Wall Model [14], Biba Integrity Model [15], Lipner Model [16] and the Class Security Model [17-18]. The authors used the Alloy formal language to define these models and the Alloy Analyzer to validate their consistency. In their work, they listed the ordered security classes (Top Secret, Secret, Confidential, and Unclassified) and their compartments (Nuclear, Technical, and Biological) and defined those using signatures. Then, the possible combinations and the relationships between classes and compartments were specified. Facts were used to set the model constraints and to prove that the model is consistent. In the Biba Integrity model, the authors listed the subject security clearance and the object security classes and then modeled the constraints of how subjects can read/write objects based on "NoReadDown" and "NoWriteUp" properties.

3   Clinical Information Systems Security Model

Security of medical records is a very important issue in clinical information systems. Security policies have to be carefully designed in order to limit the number of users that can access patient records and to control the operations that may be applied to the records themselves. Thus, it is very critical to protect confidentiality of records and their data integrity. Anderson [4] developed a policy for clinical information systems that combine confidentiality and integrity to assure patient privacy and record validity.

The policy assumes that personal health information concerns one individual at a time and is contained in a medical record. As stated in [4], the policy "principles are derived from the medical ethics of several medical societies, and from the experience and advice of practicing clinicians". It is expressed based on two sets of principles: The Access Principles set deals with confidentiality and the creation, deletion, confinement, aggregation, and Enforcements Principles set handles the integrity:

- Confidentiality principles:
  - o Access Principle 1: Each medical record has an access control list containing the individuals and groups who are able to read and append information to the record. The system must restrict access to record to those identified on the list and deny access to anyone else.
  - o Access Principle 2: One of the clinicians (responsible clinician) on the access control list must have the right to add other clinicians to the access control list.
  - o Access Principle 3: The responsible clinician must notify the patient of the names on the access control list whenever the patient's medical record is opened.
  - o Access Principle 4: The name of clinician, the date and the time of the access of medical record must be recorded. Similar information must be kept for deletions.

    Note that auditors cannot access original medical records; instead copies are provided for this purpose to prevent auditors from changing the original records. Medical records can be read and altered by the clinicians by whom the patients have consented to be treated.

- Integrity principles:
  - o Creation Principle: A clinician may open a record, with the clinician and the patient on the access control list. If the record is opened as a result of referral, the referring clinician must also be on the access control list.
  - o Deletion Principle: Clinical information cannot be deleted from a medical record until the appropriate time has passed.
  - o Confinement Principle: Information from one medical record may be appended to a different medical record if and only if the access control list of the second record is a subset of the access control list of the first.
  - o Aggregation Principle: Measures for preventing the aggregation of patient data must be effective. In particular, a patient must be notified if anyone is to be added to the access control list for the patients' record and if that person has access to a large number of medical records.
  - o Enforcement Principle: Any computer system that handles medical records must have a subsystem that enforces the preceding principles. The effectiveness of this enforcement must be subject to evaluation by independent auditors.

Therefore, based on Clinical Information Systems Security Policy highlighted rules, and assuming the existence/respect of the non-highlighted rules, the system to be implemented must ensure confidentiality by maintaining an access control lists containing users able to read and append original or copied records and grant access to other users. Moreover, the system must maintain data integrity by allowing the appending of information from one record to another if and only if the access control list of the second record is a subset of the access control list of the first. Responsible clinicians must notify patients regarding the names of clinicians on the access control list.

3.1 Clinical Information Systems Security Policy Implementation

In order to implement the model, a clinical information system is used as an example to demonstrate consistency with respect to Clinical Information Systems Security Policy. The system users can access medical records if their names are contained in access control lists attached to the medical records - one list per record. Medical records cover the personal health information of individuals. There are two versions of medical records: the original version and the copy version. Access to original copies is restricted only to clinicians and patients defined in the access control lists. Auditors are allowed to alter only the copy version of medical records. Regarding the framework used for implementation, the Alloy language and the Alloy Analyzer (based on its available features and its ability to check system consistency and to generate instances) were used for implementation.

Table 1 lists system users. There are three types of users: patients, clinicians and auditors. However, and according to Access Principle 2, the system has an additional user named the responsible clinician who has the right to add other clinicians to an

Table 1. Clinical Information System Users

Users	Description
Ptns	Patients
Clns	Clinicians
ClnsR	Responsible clinicians
Adts	Adults

Table 2. Clinical Information System Patients

Patients	Description
P1	Patient 1
P2	Patient 2

Table 3. Clinical Information System Clinicians

Clinicians	Description
C1	Clinician 1
C2	Clinician 2
C3	Clinician 3
C4	Clinician 4
C5	Clinician 5

Table 4. Clinical Information System Responsible Clinicians

Responsible clinicians	Description
CR1	Responsible Clinician 1
CR2	Responsible Clinician 2

Table 5. Clinical Information System Auditors

Auditors	Description
A1	Auditor 1
A2	Auditor 2
A3	Auditor 3

access control list. Tables 2, 3, 4 and 5 list available patients, clinicians, responsible clinicians, and auditors, respectively. Accordingly, the system has two patients (P1 and P2), five clinicians (C1, C2, C3, C4, and C5), two responsible clinicians (CR1 and CR2), and three auditors (A1, A2, and A3).

Table 6 shows two versions of medical records: the original version MRO and the copy version MRC. The system covers three medical records (mc1, mc2 and mc3) defined in table 7 and their respective copies (mo1, mo2 and mo3) shown in table 8. Each medical record contains personal health information that concerns one individual at a time.

Table 6. Clinical Information System Medical Records Versions

Medical records	Description
MRC	Medical Records Copies
MRO	Medical Records Original

Table 7. Clinical Information System Medical Records (Original)

Medical records	Description
mc1	Medical record 1
mc2	Medical record 2
mc3	Medical record 3

Table 8. Clinical Information System Medical Records (Copy)

Medical records	Description
mo1	Medical record 1
mo2	Medical record 2
mo3	Medical record 3

Table 9 determines the different access control lists used by the clinical information system. There are two types of lists: lists accessible by patients and clinicians by whom the patients have consented to be treated and they are labeled lst1, lst2 and lst3, and lists accessible by the auditors who are given access to copies of patients medical records and they are labeled lstA1, lstA2 and lstA3.

Table 9. Clinical Information System Access Control Lists

Access control lists	Description
lst1	Access control list 1 covering one patient and clinicians.
lst2	Access control list 2 covering one patient and clinicians.
lst3	Access control list 3 covering one patient and clinicians.
lstA1	Auditor Access control list covering the auditors.

Table 10 shows the domain of system access control lists. For instance, lst1 is the access control list of the medical record mo1 that concerns patient P1. In addition to P1, the clinicians C1, C4, C5 and CR1 are in lst1 and are allowed to access mo1, whereas CR1 is the responsible clinician who can add other clinicians to lst1. Similarly, lst2 is the access control list of mo2. It contains P2, C2, C3, C4 and CR2. However, list lst3 concerns mo3 of the patient P2; it contains C2 and CR2. Note that mo2 and mo3 are two different medical records for the same patient P2 assuming that the system can open a new record for a patient regardless if he/she has a previous medical record in the system, with the possibility to append the information from an existing record to the new record as per the Confinement Principle, and then discard the old record information after a certain period of time as per the Deletion Principle.

Table 11 displays a matrix showing the system relationships among objects in the clinical information system. The objects and relations are used to model and validate the clinical information system security policy. In this table and as per the Confinement Principle, the information of mo2 can be appended to mo3 since lst3 is a subset of lst2. The auditors in the system are used to enforce the system principles (Enforcement Principle). They are given access to the copies of the system medical records for checking purposes. Auditors can update these copies without any constraints because their alteration will not affect system integrity. Additionally, patients are allowed to read their medical records only, whereas clinicians are given read and append access to the medical records according to the access control lists. Moreover, responsible clinicians can add new clinicians to access control lists. They must notify patients about the names of clinicians eligible to read and append their records (i.e., clinicians available in the access control list) and also whenever a clinician is to be added to the medical record access control list (Aggregation Principle).

Table 10. Clinical Information System Access Control Lists Domain

Access control list	Medical record	Patient	Clinicians
lst1	mo1	P1	C1, C4, C5, CR1
lst2	mo2	P2	C2, C3, C4, CR2
lst3	mo3	P2	C2, CR2
lstA1	mc1, mc2, mc3	P1, P2	A1, A2, A3

Table 11. Clinical Information System Clinical System Relations

Object/ Relation	accessed by	read	append	notify	infoappendedto	addnew Clnsto
mo1	lst1				-	
mo2	lst2				mo3	
mo3	lst3				-	
mc1	lstA1					
mc2	lstA1					
mc3	lstA1					
P1		mo1	-	-		
P2		mo2, mo3	-	-		
C1		mo1	mo1	-		-
C2		mo2	mo2	-		-
C3		mo2	mo2	-		-
C4		mo1, mo2	mo1, mo2	-		-
C5		mo1	mo1	-		-
CR1		mo1	mo1	P1		lst1
CR2		mo2, mo3	mo2, mo3	P2		lst2, lst3

The clinical information system can now be represented using the Alloy language. First, the system main entities represented in the above tables are declared as shown in the following sections of code:

- Section 1 declares the main entities of the clinical information system:
    - Users: users set covering all system users.
    - HcUsrs: health care users set as part of users set.
    - Ptns: set of patients as part of users set.
    - Clns: set of clinicians as part of HcUsrs set.
    - ClnsR: set of responsible clinicians as part of Clns set.
    - Adts: set of auditors as part of users set.
    - MRs: set of all medical records defined in the system.
    - MRO: set of original medical records as part of MRs set.
    - MRC: set of medical records copies as part of MRs set.
    - ACL: set of access control lists attached to the original medical records.
    - AACL: set of auditor access control lists attached to the copy version of medical records.

```
//declaration section
abstract sig Users{ }//all users
sig HcUsrs extends Users{ }//health care users

sig Ptns extends Users{read:one MRO}//Patients
sig Clns extends HcUsrs{read:some MRO,append:some MRO}//clinicians
sig ClnsR extends Clns{notify:some Ptns, addnewClnsto:some ACL}//responsible clinicians
sig Adts extends Users{read:some MRC,append:some MRC}//auditors

abstract sig MRs{ } //all medical records
sig MRO extends MRs{accessedby:one ACL}//Medical Records original
sig MRC extends MRs{accessedby:one AACL}//Medical Records Copy

abstract sig ACL{contains: some Ptns, Contains: some Clns} //Access control lists
abstract sig AACL{contains: some Adts}//auditors access control list
```

Section 1. Clinical Information System Entities

Section 1 also defines the relations among entities. The relation "read" between Ptns and MRO means that patients can read information from their original medical records. Similarly, "read" and "append" relations between Clns and HcUsrs means that clinicians can read and append patients medical records MRO. The responsible clinicians ClnsR using the "addnewClnsto" relation are responsible to add new clinicians to access control lists and "notify" patients accordingly. Auditors are part of system users. They are allowed to alter the copy version of system medical records. Medical records set – MRO - is "accessedby" access control lists set while MRC set is accessed by auditors access control list AACL. The access control list ACL "contains" patients and clinicians; however, AACL contains only auditors.

- Section 2 defines the instances of the sets declared in section 1. Accordingly, there are three original medical records instances mo1, mo2 and mo3. Following the Confinement Principle of the Clinical Information Systems Policy, original medical records information can be appended to other medical records. Thus, the relation "infoappendedto" is defined to serve this purpose. In addition to the original medical records, the system defines three instances of copy version medical records: mc1, mc2 and mc3. Moreover, lst1, lst2 and lst3 are three instances belonging to ACL and lstA1 is one instance belonging to the AACL set. Also the system declares five clinicians: C1, C12, C3, C4 and C5, two patients: P1 and P2, three auditors: A1, A2 and A3, and two responsible clinicians: CR1 and CR2.

```
//assume a medical record concerns one individual at a time
one sig mo1,mo2,mo3 extends MRO{Infoappendedto: one|MRO} //medical orginal record instances
one sig mc1,mc2,mc3 extends MRC{ } //3 different copies of medical records

one sig lst1,lst2,lst3 extends ACL {} //access control list instances
one sig lstA1 extends AACL{} //auditors access control list instances

one sig C1,C2,C3,C4,C5 extends Clns{}//clinicians instances
one sig CR1,CR2 extends ClnsR{}//responsible clinicians instances

one sig P1,P2 extends Ptns{} //2 patients
one sig A1,A2,A3 extends Adts{}// 3 auditors
```

Section 2. Clinical Information System Instances

```
fact
{

//define content of lst1: C1, C4,C5 and CR1
lst1.contains=P1
lst1.Contains=C1+C4+C5+CR1

//define content of lst2: P2 and C2, C3, C4 and CR2
lst2.contains=P2
lst2.Contains=C2+C3+C4+CR2

//define content of lst3: P2, C2 and CR2 -> lst3 is subset of lst2
lst3.contains=P2
lst3.Contains=C2+CR2

//copy of medical record is accessed by ONE AND ONLY ONE auditors access control list only
//lstA1 contains A1, A2 and A3
lstA1.contains=A1 +A2+A3
```

Section 3. Clinical Information System Constraints (part 1)

- Section 3 highlights the beginning of the fact{} procedure where the system constraints are set. The first two lines specify explicitly the contents of the access control list lst1. List lst1 is accessed by patient P1 and clinicians C1, C4, C5 and CR1 where CR1 is the responsible clinician for lst1. List lst2 contains P2, C2, C3, C4 and CR2, and lst3 contains P2, C2 and CR2. lst3 is a subset of lst2 since P2, C2 and CR2 are also part of lst2. The last line of this section states that the auditor access control list lstA1 contains A1, A2 and A3. The symbol "+" is used for union.
- Section 4 determines that the medical records mo1, mo2 and mo3 are accessed by lst1, lst2 and lst3, respectively.

```
//orginal medical record is accessed by ONE AND ONLY ONE access control list only
mo1.accessedby=lst1
mo2.accessedby=lst2
mo3.accessedby=lst3
```

Section 4. Clinical System Constraint (part 2)

- Section 5 shows the constraints related to the read and append relations. It determines records that can be read or appended by clinicians. This section makes use of negation when there are multiple options and explicit declaration when there is only one option. For instance, patient P1 can read mo1 only. However, P2 has two records in the system mo2 and mo3. In this case, the second line states that P2 is not able to read mo1; thus, he/she can read either mo2 or mo3 at a time. The same applies to the subsequent lines of code: clinician C1 is a member of lst1 only. For this reason C1 can read/append mo1 only. However, C2 is part of lst2 and lst3. C2 cannot read/append mo1 but he/she has the option to access mo2 or mo3. Regarding the responsible clinicians CR1 and CR2, they have additional roles in the system other than reading/appending medical records. CR1 is able to add new clinician to the access control list lst1 and notify C1, accordingly. The last 4 lines of code in the section determine that CR1 notifies P1 when he/she adds new clinician to lst1 and CR2 notifies patients other than P1 (i.e., P2) upon adding new clinician to a list different than lst1 (i.e., lst2 or lst3).

Note that the section 5 does not cover any constraints regarding the copy version of medical records accessed by the auditors since their alteration will not affect system integrity.

```
//each patient can read his/her record only.
P1.read=mo1
P2.read!=mo1

C1.read=mo1
C1.append=mo1

C2.read!=mo1
C2.append!=mo1

C3.read=mo2
C3.append=mo2

C4.read=mo2
C4.append=mo2

C5.read=mo1
C5.append=mo1

CR1.read=mo1
CR1.append=mo1

CR2.read!=mo1
CR2.append!=mo1

CR1.notify=P1
CR2.notify!=P1

CR1.addnewClnsto=lst1
CR2.addnewClnsto!=lst1
```

Section 5. Clinical Information System Constraints (part 3)

3.2   Clinical Information System Security Policy and Alloy Analysis

Figure 1 displays the clinical information system meta model generated by the Alloy system. It shows multiple users of the system: auditors - Adts, health care users - HcUsrs, patients – Ptns, and clinicians – Clns, as part of health care users. Additionally, the figure displays the system access control lists – ACL, and auditors' access control lists – AACL, and their instances. Moreover, the model shows two types of medical records: original version of medical records (MRO) and the copy version (MRC) and their instances. However, the meta model does not show any constraints. Executing the system using the Alloy Analyzer will generate instances based on defined constraints.

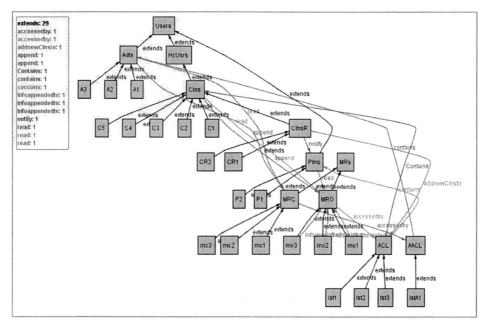

Fig. 1. The Clinical Information System Meta Model

Testing system consistency is done by running the system predicates and generating possible instances then validating them. In order to test the constraints specified in the fact procedure, we need to write a predicate and run it.

- Section 6 declares an empty predicate used to test the system consistency based on the defined facts. Executing the example() will produce the output specified in figure 2. The figure shows that an "instance found" and "Predicate is consistent". It takes around 78ms to determine consistency and find an instance.

```
// run example to show consistencies/ inconsistencies
pred example( ){
}
run example
```

Section 6. Clinical Information System Predicate

```
Executing "Run example"
   Solver=sat4j Bitwidth=4 MaxSeq=4 SkolemDepth=1 Symmetry=20
   292 vars. 124 primary vars. 291 clauses. 78ms.
   Instance found. Predicate is consistent. 78ms.
```

Fig. 2. Clinical Information System Consistent Alloy Analyzer Output

Clicking the link "**Instance**" will show figure 3. More instances can be generated by pressing the "Next" button located at the top of the screen of figure 3.

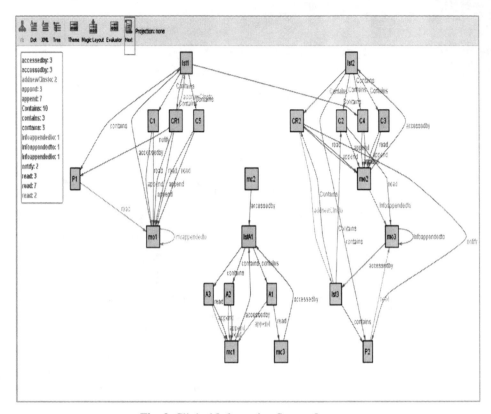

Fig. 3. Clinical Information System Instance

The generated instances demonstrate the consistency of the system as shown in tables 12- 16.

Table 12. Clinical Information System Consistency Checking (part 1)

Object/Relation	contains	consistent
lst1	P1, C1, CR1, C5, C4	Yes
lst2	CR2, C2, C4, C3, P2	Yes
lst3	P2, C2, CR2	Yes

Table 13. Clinical Information System Consistency Checking (part 2)

Object/Relation	accessedby	consistent
mo1	lst1	Yes
mo2	lst2	Yes
mo3	lst3	Yes
mc1	lstA1	Yes
mc2	lstA1	Yes
mc3	lstA1	Yes

Table 14. Clinical Information System Consistency Checking (part 3)

Object/Relation	read	append	consistent
P1	mo1	-	Yes
P2	mo3	-	Yes
C1	mo1	mo1	Yes
C2	mo3	mo2	Yes
C3	mo2	mo2	Yes
C4	mo2	mo2	Yes
C5	mo1	mo1	Yes
CR1	mo1	mo1	Yes
CR2	mo2	mo2	Yes
A1	mc3	mc1	Yes
A2	mc1	mc1	Yes
A3	mc1	mc1	Yes

Table 15. Clinical Information System Consistency Checking (part 4)

Object/Relation	infoappendedto	consistent
mo1	mo1	Yes
mo2	mo3	Yes
mo3	mo3	Yes

Table 16. Clinical Information System Consistency Checking (part 5)

Object/Relation	addnewClnsto	notify	consistent
CR2	lst3	P2	Yes
CR1	lst1	P1	Yes

However, specifying a wrong predicate such as stating that lst1 contains the patient P2 (section 7, line 2) will cause inconsistency and the result of running the example() is displayed in figure 4.

```
// run example to show inconsistencies
pred example( ){

lst1.contains=P2 //inconsistent bcz lst1 contains only one patient
//P1.append=mo2 // inconsistent bcz patient cannot append medical record

}
run example
```

Section 7. Clinical Information System Inconsistent Predicate

```
Executing "Run example"
  Solver=sat4j Bitwidth=4 MaxSeq=4 SkolemDepth=1 Symmetry=20
  293 vars. 124 primary vars. 294 clauses. 47ms.
  No instance found. Predicate may be inconsistent. 0ms.
```

Fig. 4. Clinical Information System Alloy Analyzer Inconsistent Output

Thus, the clinical information system adopting the Clinical Information Security Policy is a consistent system. Any misbehavior will results in an inconsistency where no instances are found.

4   Conclusion

In this paper, we presented the Clinical Information System Security Policy. We used system examples based on the defined security model. We formalized the system according to the model then checked its consistency. Since Alloy allows expressing systems as set of logical constraints in a logical language based on standard first-order logic, we used it to define the system and policy. When creating the model, we specified the system users and subjects then Alloy compiled a Boolean matrix for the constraints, and we asked it to check if a model is valid, or if there are counter examples.

Acknowledgements. This work was funded by the Lebanese American University.

References

[1] Summers, C.: Computer Security: Threats and Safeguards. McGraw Hill, New York (1997) ISBN-13: 978-0070694194

[2] Jackson, D.: Alloy 3.0 Reference Manual (2004), http://alloy.mit.edu/reference-manual.pdf (retrieved on September 24, 2012)

[3] Seater, R., Dennis, G.: Tutorial for Alloy Analyzer 4.0 (2011), http://alloy.mit.edu/tutorial4 (retrieved on September 24, 2012)

[4] Anderson, R.: A Security Policy Model for Clinical Information Systems. In: Proceedings of the 1996 IEEE Symposium and Security and Privacy, pp. 30–43. IEEE Press, Oakland (1996)

[5] Hassan, W., Logrippo, L.: Detecting Inconsistencies of Mixed Secrecy Models and Business Policies. University of Ottawa, Canada, Technical Report (2009)

[6] Bell, L., LaPadula, E.: Secure Computer Systems: Mathematical Foundations. Technical Report 2547, Volume I. The MITRE Corporation (1976)

[7] Ferraiolo, D.F., Kuhn, D.R.: Role-Based Access Control. In: Proceedings of the 15th National Computer Security Conference, Baltimore, MD, USA, pp. 554–563 (1992)

[8] Viega, J., Evans, D.: Separation of Concerns for Security. In: Proceedings of the Workshop on Multi-Dimensional Separation of Concerns in Software Engineering, Limerick, Ireland, pp. 126–129 (2000)

[9] Zao, J., Hoetech, W., Chu, J., Jackson, D.: RBAC Schema Verification using Lightweight Formal Model and Constraint Analysis. In: Proceedings of 8th ACM Symposium on Access Control Models and Technologies, Boston, MA, USA (2003)

[10] Hassan, W., Logrippo, L., Mankai, M.: Validating Access Control Policies with Alloy. In: Proceedings of the Workshop on Practice and Theory of Access Control Technologies, Quebec, Canada (2005)

[11] Shaffer, A., Auguston, M., Irvine, C., Levin, T.: A Security Domain Model to Assess Software for Exploitable Covert Channels. In: Proceedings of the ACM SIGPLAN Third Workshop on Programming Languages and Analysis for Security, Tucson, Arizona, USA, pp. 45–56 (2008)

[12] Misic, J., Misic, V.: Implementation of Security Policy for Clinical Information Systems over Wireless Sensor Networks. Ad Hoc Networks Journal 5, 134–144 (2006)

[13] Haraty, R.A., Boss, N.: Modeling and Validating Confidentiality, Integrity, and Object Oriented Policies using Alloy. In: Security & Privacy Preserving in Social Networks. Springer (2013) ISBN 978-3-7091-0893-2

[14] Brewer, D., Nash, M.: The Chinese Wall Security Policy. In: Proceedings of the IEEE Symposium on Research in Security and Privacy, Oakland, CA, USA, pp. 206–214 (1989)

[15] Biba, K.J.: Integrity Considerations for Secure Computer Systems. Technical Report MTR-3153. The MITRE Corporation (1977)

[16] Lipner, S.B.: Non-discretionary Controls for Commercial Applications. In: Proceedings of the IEEE Symposium on Security and Privacy, Oakland, CA, USA, pp. 2–10 (1982)

[17] Haraty, R.A.: A Security Policy Manager for Multilevel Secure Object Oriented Database Management Systems. In: Proceedings of the International Conference on Applied Modeling and Simulation, Cairns - Queensland, Australia (1999)

[18] Haraty, R.A.: C2 Secure Database Management Systems – A Comparative Study. In: Proceedings of the Symposium on Applied Computing. San Antonio, TX, USA, pp. 216–220 (1999)

Applying a BP Neural Network Model to Predict the Length of Hospital Stay

Jing-Song Li[1,*], Yu Tian[1], Yan-Feng Liu[1], Ting Shu[2], and Ming-Hui Liang[2]

[1] Zhejiang University, Heathcare Informatics Engineering Research Center, Hangzhou, China
ljs@zju.edu.cn, {ty.1987823,bmelyf}@gmail.com
[2] MOH of China, National Institute of Hospital Administration, Beijing, China
Nctingting@126.com, Liang2002@vip.sina.com

Abstract. Length of hospital stay (LOS) is closely related to the control of medical costs and the management of hospital resources. In this study, we implemented a data mining approach based on Back-Propagation (BP) neural net-works to construct a LOS prediction model that can help doctors and nurses individualize patient treatment. We analyzed medical data from 921 patients whowere diagnosed as cholecystitis and treated in a Chinese hospital between 2003and 2007. Our prediction model achieved approximately 80% accuracy, and revealed 5 LOS predictors: days before operation, wound grade, operation approach, charge type and number of admissions. The model can be easily used toprovide suggestions for doctors and nurses determining patient LOS.

Keywords: Length of hospital stay, LOS prediction, BP neural network, Data mining.

1 Introduction

With the continuous development of medicine and healthcare technology, the level of medical service is rising. At the same time, the excessive growth of medical costs puts significant pressure on governments and their healthcare service organizations. The effective control of medical costs and expenses has also become a priority issue for all healthcare institutions. The ideal medical service pattern should allocate medical resources accurately based on the understanding of the patient's condition and needs to maximize the efficiency of medical services. Study and analysis of the patient's condition, as well as of the medical process management, are needed to achieve this ideal pattern of medical service, and length of stay in the hospital (LOS) is an important variable to consider in this research.

Previous studies have used LOS to assess medical costs and effectiveness and have made use of many methods to estimate or predict the LOS of particular patients. LOS has a direct link with medical costs and resource consumption. Stephen Martin[1]suggested that, in the absence of medical cost data, patient LOS should be used as the principal benchmark of not only the efficiency of inpatient medical treatmentbut

---

* Corresponding author.

G. Huang et al. (Eds.): HIS 2013, LNCS 7798, pp. 18–29, 2013.
© Springer-Verlag Berlin Heidelberg 2013

also of the implied medical resource consumption. Samir M Fakhry[2] showed that adult trauma patient LOS correlates closely with trauma center profitability. Patient LOS also plays an important role in the assessment of medical charge policy; several studies have used LOS to measure the consequence of given medical charge policy [1-6], and LOS has been shown to be an effective indicator in the assessment of disease treatments and risk factors. Focusing on one disease, researchers identified factors that may cause a longer LOS, evaluated the effect of alternative treatments on LOS and assessed disease risk factors [7-17].

Previous studies have used various statistical methods to assess and predict LOS, including multiple linear regression and logistic regression [18-24]. Although these statistical methods have been applied widely, their ability to predict LOS has been unsatisfactory. Other modeling techniques have achieved better LOS prediction performance [11, 15-17, 23, 25-27]. Geert Meyfrodt [11] showed that a Gaussian process model demonstrated significantly better discriminative power than theEuroS-CORE Yang chin sheng[15] etc. used a Support Vector Machine (SVM) model to predict LOS for burn patients. They emphasized that traditional linear regression may not have sufficient flexibility or robustness across different applications because it explicitly enforces linear relationships among independent variables and requires strict conformance to specific and often rigid assumptions, such as the assumption that residuals are independent and identically distributed. Therefore, artificial intelligence techniques may have some strong advantages over traditional statistical models in LOS prediction.

Data mining is an artificial-intelligence-powered tool that can be used to discover useful information within a database that can then be used to improve actions. Data mining can involve the analysis of observational data sets to reveal unsuspected relationships among multiple variables and summarizing the data in novel ways that can be both understandable and useful to the data owner. In medical practice, a variety of variables need to be considered when determining hospital LOS, such as demographic factors, characteristics of the disease, treatment modality, etc. A promising method for identifying potential LOS predictors would be to use data mining on a large dataset gathered from the Health Information System (HIS) and Electric Medical Records (EMR).

In this study, we apply a data mining approach based on BP neural network to construct a prediction model for LOS. With the help of the data mining approach, we are able to obtain the LOS predictors and their combined mode for patients who underwent surgery for cholecystitis. Based on different combined modes, we employ a BP neural network to construct LOS prediction models. Then, we compare different LOS prediction models and identify the reasonable predictors.

2   Methods

2.1   Patients and Data Collection

The data we mined and analyzed were selected from the Oracle database of the HIS of a large hospital in Zhejiang Province, China, which contains over 700 beds and treats over 15 thousand inpatients every year. We collected data from a total of 921 hospitalized cholecystitis patients who were treated in the hospital between 2003 and 2007. The raw medical data in the database were voluminous and heterogeneous. The medical data recorded in the HIS included various images, interviews with patients, laboratory data, and the physician's observations and interpretations. We selected variables of interest from the Oracle database with the help of doctors that were familiar with cholecystitis. These variables included demographic factors such as age, gender, and patient admission condition; operation information, including approach, number, scale, anesthesia and wound grade; and hospitalization information including charge type, days before operation and number of admissions. Table 1 shows the distribution of categorical variables in the data set. To express the impact of different age groups on the LOS, we divided all patients into three age groups. The "patient admission condition" expresses the patient's disease situation when they were admitted to the hospital. It consists of three levels: level 1 indicates that the patient's condition was critical and had to be treated immediately; level 2 indicates that the patient's condition was an emergency, while level 3 means the patient's condition was normal. There were three possible wound grade levels: Level I (aseptic), Level II (contaminated) and Level III (infected). The charge type reflects the mode of payment used by the patient, which could be one of four different possible types: type 1, labor insurance, means medical costs are borne entirely by the patient's employer; type 2,

Table 1. Statistial results for categorical variables

Variable	Detail	Occurrence (%)	Number of records
Gender	Male	33.37	307
	Female	66.63	618
Age	Age<45	31.41	289
	45<Age<57	34.24	315
	Age>57	34.54	316
Patient admission condition	Level 2	10.43	96
	Level 3	89.57	824
Operation approach	Cholecystectomy	15.43	142
	Laparoscopic cholecystectomy	84.57	778
Operation scale	Large	95	874
	Medium	4.35	40
Anesthesia	General anesthesia	98.26	904
	Local anesthesia	1.63	15
Wound grade	Level I	14.57	134
	Level II	83.26	766
	Level III	0.87	8
Charge type	Type 1	47.93	441
	Type 2	15.98	147
	Type 3	31.74	292
	Type 4	4.35	40

medical insurance, means medical costs are borne partly by the patient while the rest is shared by the Government and the patient's employer; type 3, self-pay, means all medical costs are the patient's responsibility; and type 4 covers other payment modes that have little relevance to our study. Table 2 contains the breakdown of numeric variables in the data set. "Operation number" represents the number of operations that patients underwent during their hospitalization, while number of admissions reflects the number of times patients were hospitalized; most of the patients in our study were hospitalized once and underwent one operation. Days before operation are a reflection of patients' physiological condition and illness. LOS was our dependent variable.

Table 2. Statistical results for numeric variables

Variable	Description			
	Min	Max	Mean	SD
Operation number	1 time	3 times	1.03 times	0.178
Number of admissions	1 time	6 times	1.068 times	0.334
Days before operation	0 day	50 days	2.487 days	3.302
LOS of inpatients	2 days	74 days	7.066 days	5.218

2.2   Medical Data Mining Approach

Data mining is an artificial-intelligence-powered tool that can be used to discover useful information within a database, which can then be used to improve actions. Data mining can be applied to the analysis of observational data sets to find unsuspected relationships among multiple variables and to summarize the data in novel ways that can be both understandable and useful to the data owner. When mining medical data, we are not only concerned with generating and evaluating mathematical models; we are also sensitive to concerns regarding data privacy and must take the heterogeneity of the data into consideration[28]. For this study, we adopted the Cross Industry Standard Process for Data Mining (CRISP-DM). The CRISP-DM project was conceived in 1996 by three "veterans" of the immature DM market: Daimler-Chrysler (then Daimler-Benz), SPSS (then ISL) and NCR. The project was defined and validated as an industry- and tool-neutral DM process model, which is now the industry standard methodology for DM and predictive analytics across a wide range of sectors. The CRISP-DM process has been found useful for medical data analysis. Eduardo Rivo found the CRISP-DM process to be very suitable for lung cancer surgery analysis and showed that it improved medical care decision making as well as knowledge and quality management[29]. In our study, we focused on data understanding, data preparation, modeling and evaluation.

2.3   BP Neural Network

We applied a BP neural network as the modeling technique in our data mining process. There are various classification and prediction algorithms, such as decision trees, Bayesian belief networks, QUEST, Artificial Neural Networks (ANN), CHAID and so on. ANNs are based on brain structure and, similar to the brain, ANNs can recognize patterns, manage data and learn. They consist of artificial neurons that implement the essential functions of biological neurons. Back-Propagation is one technique used to optimize ANN architecture. A Back-Propagation Network is a multi-layered feed-forward neural network, and back-propagation involves giving an input vector to the network, comparing the desired output to the network output for the given input vector, and changing each neural weight by an amount equal to the derivative of the error with respect to each weight multiplied by some learning rate.

Daniel J. Sargent [27] has previously published research that compared Artificial Neural Networks with other statistical approaches. We chose a BP neural networks for data mining because these networks learn by example; consequently, the details of how to forecast the categories of LOS are not needed. What is needed is a collection of records that is representative of all of the variations of hospital LOS. As one of the architectures of ANNs, BP neural networks are also advantageous because they allow arbitrary nonlinear relations between independent and dependent variables.

2.4 Empirical Study

We chose Clementine 12.0 as our medical data mining tool because it is a commercial tool (SPSS, 2009) that conforms to the CRISP-DM process model. Our research process is illustrated in Fig. 1 and included the following steps, which will be described in more detail below: data collection, data preparation, correlation analysis and BP neural network training.

We collected data from a total of 921 hospitalized cholecystitis patients who were treated in the hospital between 2003 and 2007.

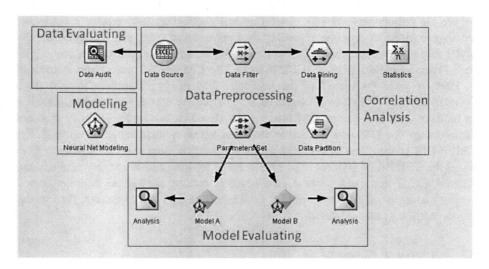

Fig. 1. Data analysis procedure implemented in Clementine

To prepare our data for analysis, we deleted or de-identified variables that related to privacy and security, such as name and address, to protect the anonymity of the patients. We divided the dataset into three categories or bins (shown in Table 3) and then predicted the LOS for each bin.

Table 3. Classification of the number of the stay in hospital

Categories	Number of days	Number of records	Mean Value of LOS
1	2<=t<5	242	3
2	5<=t<7	346	5
3	7<=t<74	332	9

For our correlation analysis, we assumed that all input and output variables were independent. The correlation strength was assessed using chi-square statistics and a significance level (α) of 0.05. We divided our significant results into three correlation strengths: strong (1-p>0.95), medium (0.90<1-p<0.95) and weak (1-p<0.90), where p was the test probability value. At the same time, we also calculated the correlation coefficient for every variable.

To train the BP neural network, we had to set the value of several critical neural network parameters. Half of the records were selected randomly as the training set, while the other records were set aside as the test set. We chose to have a single hidden layer, 3 input neurons, 15 hidden neurons and 3 output neurons. We used an initial learning rate η of 0.35 and an η decay rate of 0.1. The learning rate was adjusted adaptively, and the initial the value was 0.9, according to experience. For Model A, we used all of the variables in Tables 1 and 2 (except LOS) as inputs and adopted LOS as our output variable. For Model B, we limited our inputs to variables that were strongly correlated with LOS.

Finally, we used the test set to assess the prediction performance and model sensitivity of both models.

3 Results

3.1 Correlation Analysis Results

We assumed that all input and output variables were independent. The correlation strength was assessed using chi-square statistics with a significance level (α) of 0.05. We divided our significant results into three correlation strengths: strong (1-p>0.95), medium (0.90<1-p<0.95) and weak (1-p<0.90), where p is the test probability value. We also calculated the correlation coefficient for every variable with LOS. The results are shown in Table 4.

As can be seen from Table 4, all factors except Operation scale, Anesthesia and Gender showed medium to strong correlations with LOS. Therefore, we speculated that the patient's gender and the type of anesthesia used had little effect on LOS, and we did not consider them in our BP neural network modeling. The correlation

Table 4. Correlation strength of all the variables

Variables	Correlation Strength	Correlation
Number of admissions	0.111	Strong
Operation scale	0.035	Weak
Operation number	0.189	Strong
Days before operation	0.388	Strong
Patient admission condition	-0.125	Strong
Operation approach	-0.484	Strong
Wound grade	0.064	Medium
Anesthesia	-0.049	Weak
Gender	-0.044	Weak
Charge type	-0.116	Strong
Age	0.172	Strong

analysis also indicated that all of the variables except Patient admission condition and Operation approach were positively correlated with LOS; however, the sign of the correlation between a variable and a result was determined by the way the data were transformed and interpreted. We were more concerned with the absolute values of the correlation coefficients. Operation approach, for instance, included two different approaches, cholecystectomy and laparoscopic cholecystectomy, according to ICD9CM standard.

3.2   Prediction Model Evaluation Results

We built two models using different input vectors; one included all variables (Model A), and the other only included factors that had a strong correlation with LOS (Model B). The performance of the models is shown in Table 5. We could see that the precision of Model A was only 3-4% higher than Model B, suggesting that correlation analysis was important at the input layer of the neural network.

Table 5. Results of the analysis to different Model

	Model A	Model B
Correct	83.19%	79.2%
	381	362
Wrong	16.81%	20.8%
	77	96
Total	458	458

We evaluated the model quality using measures of sensitivity, specificity, precision, false positives and false negatives. Sensitivity, which measures the fraction of positive cases that are classified as positive, and specificity, which measures the fraction of negative cases that are classified as negative, are the two most frequently used measures for medical models. [30] We classified the LOS of patients with cholecystitis into three categories. Table 6 shows the prediction performance of Model A. Theaverage confidence level of correctly classified records (mean correct) for Model A was 0.758, while the average confidence level of incorrectly classified records (mean incorrect) was 0.446. Table 7 shows the predictive results of Model B. The mean correct value of Model B was 0.71, while the mean incorrect value was 0.395.

Table 6. The results of Model A evaluation measures

Evaluation Measures	Model A		
	Category 1	Category 2	Category3
Sensitivity	0.8430	0.8121	0.8615
Specificity	0.9071	0.8711	0.9796
Precision	0.2491	0.3598	0.3318
False positives	0.1570	0.1289	0.1385
False negatives	0.0929	0.1879	0.0204

Table 7. The results of Model B evaluation measures

Evaluation Measures	Model B		
	Category 1	Category 2	Category3
Sensitivity	0.8471	0.7659	0.8193
Specificity	0.8776	0.8519	0.9830
Precision	0.2563	0.3515	0.3200
False positives	0.1529	0.2341	0.1807
False negatives	0.1224	0.1481	0.0170

To enhance the interpretability of our neural network prediction models, we analyzed the importance of model inputs, and the results of these analyses are shown in Figs. 2 and 3. The input variables are listed in order of importance, from most important to least important. The value listed for each input is a measure of its relative importance, between 0 (a field that has no effect on the prediction) and 1.0 (a field that completely determines the prediction).

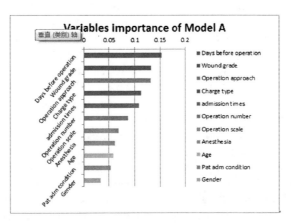

Fig. 2. Relative importance of Model A inputs

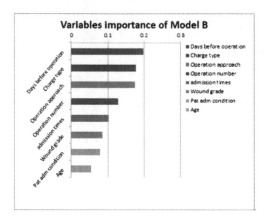

Fig. 3. Relative importance of Model B inputs

4   Discussion

In this study, we adopted data mining approach based on a BP neural network and generated two LOS prediction models using different input variables. Model A used all of the variables that we selected as inputs, and Model B used a subset of these variables as inputs, which were filtered by correlation analysis. Based on the results that we described above, Model A had a slightly higher predictive accuracy than Model B, but the performance of the two models was similar. Careful consideration of input variables is needed when using BP neural networks to predict LOS. Using correlation analysis to filter our inputs, we observed a slight decline in the prediction accuracy of the model, suggesting that BP neural networks can express some hidden relationships between input and output variables. This hidden ability could be of great importance when mining medical data and constructing prediction models.

To compensate for the lack of explanatory power of our prediction models, we analyzed the importance of variables used as input to our models. Comparing the two models, we found that 5 factors were important in predicting LOS for patients with cholecystitis. They were Days before operation, Wound grade, Operation approach, Charge type, and Number of admissions. Two important factors may contribute to the significance of days before operation in predicting the LOS. First one, it's obviously that the more days a patient is in the hospital before his/her surgery, the longer his/her LOS will be. In addition, the conditions of patients who were in the hospital for longer times prior to surgery were generally very complex or bad, which usually meant the patients also needed more time in the hospital for observation and nursing postsurgery. Wound grade reflects the extent that patients suffer from the trauma and surgery. As for the approach of the surgeon, we found that most of the patients whounderwent laparoscopic cholecystectomy were classified in the shortest LOS bin, suggesting that laparoscopic cholecystectomy reduces trauma and scarring and promotes a faster recovery. The abdominal muscle damage was reduced, and patients could ambulate and recover gastrointestinal peristalsis sooner. However, this relationship did not hold for cholecystectomy.

The charge type for cholecystectomy included normal medical insurance, labor insurance, self-payment and others. The physicians for these patients might have suggested that the patients stay more days in the hospital because doctors are evaluated by patient turnover in many hospitals in China. These patients may also have been willing to stay in the hospital longer as a health precaution because the extra time cost them little given their insurance plans, potentially wasting medical resources. The rank of the rest of the variables was lower because they simply contributed to the most important factors that we have already discussed.

In this study, we found that data mining using a BP neural network has great flexibility and robustness in predicting LOS for patients with cholecystitis. The majority of previous work has concentrated on traditional statistical methods, such as regression (Multiple linear regression, logistic regression and so on) to build LOS prediction models. Those methods can analyze existing data from a statistical perspective and predict possible LOS trends with the help of regression analysis, but they have several shortcomings: these statistical models strictly adhere to the statistical hypothesis, and the relationships among variables are constrained to linear relationships, which excludes the possibility of nonlinear relationships among the variables. In addition, these models are always used in retrospective studies; conclusions can sometimes be drawn from these models, but the conclusions are not useful for individualizing patient treatment. To overcome the shortcomings of statistical methods in predicting LOS, some researchers have implemented artificial intelligence methods. SVMs and model-tree-based regression have been shown to perform well in model training and model explanation [15], but their application adaptability and prediction accuracy need to be improved. In this study, we used CRISP-DM and BP neural network modeling to establish a superior framework covering all aspects of LOS prediction, from collection of data to model deployment. Additionally, our prediction model has greater flexibility and robustness because it uses a BP neural network. We are able to capture hidden nonlinear relationships between input variables and LOS, and we can use this model without prior knowledge of the complexity of the underlying relationships. Third, our prediction model has strong portability; our prediction model can adapt to different applications through relearning and revising. The current study focused on only one medical institution, but the data mining framework can be transported to other institutions, and it could potentially even be extended to the regional medical information system.

We not only consider the establishment of an LOS prediction model but also the application of our model. The LOS prediction model can be used as a second opinion in formulating patients' postoperative treatment plans. Doctors subjectively determine patients' hospitalization. Typically, they tend to recommend a longer LOS to ensure that patients have recovered and meet discharge indicators, but a longer LOS increases medical costs. Our model shortens hospitalization as much as possible while ensuring the rehabilitation of patients. The LOS prediction model can also be used by nurses. The prediction model can help nurses identify patients who will have to stay in the hospital longer early in their hospitalization and enable nurses to give them the specialized care they will need. Some studies have shown that careful medical attention given early on to patients who have a tendency to stay in the hospital longer can significantly reduce their LOS and their medical expenses. The LOS prediction model can also be used by patients and their families to learn their hospitalization and communicate with their doctors in a comprehensible and effective manner.

5   Conclusion

We generated a LOS prediction model for patients with cholecystitis using BP neural networks. The prediction accuracy of this model was better than 80%, and we identified 5 important predictors (Days before operation, Wound grade, Operation approach, Charge type, Number of Admissions) for the LOS of patients with cholecystitis.

References

1. Stephen, M., Peter, S.: Explaining variations in inpatient length of stay in the National Health Service. Journal of Health Economics 15, 279–304 (1996)
2. Fakhry, S.M., Couillard, D., Liddy, C.T., Adams, D., Norcross, E.D.: Trauma Center Finances and Length of Stay: Identifying a Profitability Inflection Point. Journal of the American College of Surgeons 210, 817–821 (2010)
3. Liu, Z., Dowb, W.H., Nortonb, E.C.: Effect of drive-through delivery laws on postpartum length of stay and hospital charges. Journal of Health Economics 23, 129–155 (2004)
4. Evans, J.H., Hwang, Y., Nagarajan, N.J.: Management control and hospital cost reduction: additional evidence. Journal of Accounting and Public Policy 20, 73–88 (2001)
5. Theurl, E., Winner, H.: The impact of hospital financing on the length of stay: Evidence from Austria. Health Policy 82, 375–389 (2007)
6. Taheri, P.A., Butz, D.A., Greenfield, L.J.: Length of stay has minimal impact on the cost of hospital admission. Journal of the American College of Surgeons 191, 123–130 (2000)
7. Chertow, G.M., Burdick, E., Honour, M., Bonventre, J.V., Bates, D.W.: Acute Kidney Injury, Mortality, Length of Stay, and Costs in Hospitalized Patients. Journal of the American Society of Nephrology 16, 3365–3370 (2005)
8. Nawata, K., Ii, M., Ishiguro, A., Kawabuchi, K.: An analysis of the length of hospital stay for cataract patients in Japan using the discrete-type proportional hazard model. Mathematics and Computers in Simulation 79, 2889–2896 (2009)
9. Imai, H., Hosomi, J., Nakao, H., Tsukino, H., Katoh, T., Itoh, T., Yoshida, T.: Characteristics of psychiatric hospitals associated with length of stay in Japan. Health Policy 74, 115–121 (2005)
10. Uzzo, R.D., Wei, J.T., Hafez, K., Kay, R.: Comparison of direct Hospital costs and length of stay for radical nephrectomy versus nephronsparing surgery in the managment of localized renal cell carcinoma. Adult Urology 54, 994–998 (1999)
11. Meyfroidt, G., Guiza, F., Cottem, D., Becker, W.D., Loon, K.V., Aerts, J.M., Berckmans, D., Ramon, J., Bruynooghe, M., Van den Berghe, G.: Computerized prediction of intensive care unit discharge after cardiac surgery: development and validation of a Gaussian processes model. Bmc Medical Informatics and Decision Making 11, 64–77 (2011)
12. Hung, W.J., Lin, L.P., Wub, C.L., Lin, J.D.: Cost of hospitalization and length of stay in people with Down syndrome: Evidence from a national hospital discharge claims database. Research in Developmental Disabilities 32, 1709–1713 (2011)

13. Paul, S.D., Eagle, K.A., Guidry, U.: Do Gender-based differences in presentation and management influence predictors of hospitalization costs and length of stay after an acute myocardial infarction? The American Journal of Cardiology 76, 1122–1125 (1995)
14. Furlanetto, L.M., da Silva, R.V., Bueno, J.R.: The impact of psychiatric comorbidity on length of stay of medical inpatients. General Hospital Psychiatry 25, 14–19 (2003)
15. Yang, C.S., Wei, C.P., Yuan, C.C., Schoung, J.Y.: Predicting the length of hospital stay of burn patients: Comparisons of prediction accuracy among different clinical stages. Decision Support Systems 50, 325–335 (2010)
16. Negassa, A., Monrad, E.S.: Prediction of length of stay following elective percutaneous coronary intervention. ISRN Surg. 2011, 6 pages (2011)
17. Rowan, M., Ryan, T., Hegarty, F., O'Hare, N.: The use of artificial neural networks to stratify the length of stay of cardiac patients based on preoperative and initial postoperative factors. Artificial Intelligence in Medicine 40, 211–221 (2007)
18. Lin, C.L., Lin, P.H., Chou, L.W., Lan, S.J., Meng, N.H., Lo, S.F., Wu, H.-D.I.: Model-based Prediction of Length of Stay for Rehabilitating Stroke Patients. Journal of the Formosan Medical Association 108, 653–662 (2009)
19. Houdenhoven, M.V., Nguyen, D.T., Eijkemans, M., Steyerberg, E., Tilanus, H., Gommers, D., Wullink, G., Bakker, J., Kazemier, G.: Optimizing intensive care capacity using individual length-of-stay prediction models. Critical Care 11, 1–10 (2007)
20. Ettema, R.G., Peelen, L.M., Schuurmans, M.J., Nierich, A.P., Kalkman, C.J., Moons, K.G.: Prediction models for prolonged intensive care unit stay after cardiac surgery: Systematic review and validation study. Circulation 122, 682–689 (2010)
21. Paterson, R., MacLeod, D.C., Thetford, D., Beattie, A., Graham, C., Lam, S., Bell, D.: Prediction of in-hospital mortality and length of stay using an early warning scoring system: clinical audit. Clinical Medicine, Journal of the Royal College of Physicians 6, 281–284 (2006)
22. Chang, K.C., Tseng, M.C., Weng, H.H., Lin, Y.H., Liou, C.W., Tan, T.Y.: Prediction of length of stay of first-ever ischemic stroke. Stroke 33, 2670–2674 (2002)
23. Kramer, A., Zimmerman, J.: A predictive model for the early identification of patients at risk for a prolonged intensive care unit length of stay. Bmc Medical Informatics and Decision Making 10, 1–16 (2010)
24. Giakoumidakis, K., Baltopoulos, G.I., Charitos, C., Patelarou, E., Galanis, P., Brokalaki, H.: Risk factors for prolonged stay in cardiac surgery intensive care units. Nursing in Critical Care 16, 243–251 (2011)
25. Walsh, P., Cunningham, P., Rothenberg, S.J., O'Doherty, S., Hoey, H., Healy, R.: An Artificial Neural Network Ensemble to Predict Disposition and length of stay in children presenting with bronchiolitis. European Journal of Emergency Medicine 11, 259–264 (2004)
26. Zhong, W., Chow, R., He, J.Y.: Clinical charge profiles prediction for patients diagnosed with chronic diseases using Multi-level Support Vector Machine. Expert Systems with Applications 39, 1474–1483 (2012)
27. Sargent, D.J.: Comparison of artificial neural networks with other statistical approaches: results from medical data sets. Cancer 91, 1636–1642 (2001)

Emergency Mobile Access to Personal Health Records Stored on an Untrusted Cloud

Feras Aljumah, Raymond Hei Man Leung, Makan Pourzandi, and Mourad Debbabi

Concordia University, Montreal, Qc, Canada
{f_aljum,h_leung,makan.pourzandi,debabbi}@encs.concordia.ca

Abstract. When storing files on an untrusted cloud, attribute based encryption is the cryptosystem usually chosen to securely encrypt the files while allowing fine grained access. When storing Personal Health Records (PHR), we find that allowing access to a users emergency medical records (EMRs) during an emergency would be difficult to achieve while ensuring privacy preservation. Providing ECPs with unlimited and unrestricted access to EMRs is not an acceptable solution for a privacy view point. In this work our aim is to allow ECPs the ability to access a patients EMRs, but only in the case of an emergency, preventing them from abusing their privileges. We propose a solution that solves this problem without requiring the participation of the patient in the process.

1 Introduction

As patients our medical records are stored in many locations; hospitals, pharmacies, clinics, medical labs, etc. This makes it difficult for a doctor to get the complete medical history for a patient. Personal health records (PHR) have become popular. In such a system, patients' data would be gathered in one location and is accessible to the patients.

With cloud storage we usually expect the data on the cloud to be protected by trusted servers. However, if an attacker were able to penetrate the security of the authentication server on such a cloud, he would be able to access all patient records. Since medical health records are sensitive, we assume the data is stored on an untrusted server.

Although this might seem unsecure, each file in the system is to be encrypted using an attribute based cryptosystem which would only allow users with the right credentials to decrypt the files. To better understand what we mean by attributes we could think of them as roles or privileges which that entity has. In a cloud based patient centric PHR management system, the patients have control over their medical records. When we assume the cloud to be untrusted, then this means that the files are to be encrypted with the patient's private key and access control is to be controlled by the patient.

Before a user encrypts a medical record, he associates that record with a set of attributes which describe the users who should have access to the file.

G. Huang et al. (Eds.): HIS 2013, LNCS 7798, pp. 30–41, 2013.

When a record is requested from the cloud, the user would be able to decrypt the record if and only if his key has all the attributes associated with the encrypted record.

Our proposed solution solves the problem of granting an ECP access to a patient's EMRs in case of an emergency while preventing them from abusing that power to access patients' records without an emergency.

1.1 Problem Statement

In case of an emergency, the patient's EMRs must be available to ECPs. EMRs contain basic information such as the patient's blood type, terminal illnesses, emergency contacts, etc. Granting access to these files may be easily done by adding an attribute to represent an ECP. Also, when the patients encrypt their emergency files, they must set the emergency access attribute. This solution would easily allow ECP to access any patient's emergency medical records. However, it does give too much power to ECPs. An employee at an ECP would have the power to access all patient emergency files in the system without being detected or stopped since we're assuming the cloud to be untrusted.

In our research, our goal is to come up with a method that would enable us to allow ECPs access to a patient's emergency medical records without giving them the ability to view every medical record in the system.

This paper is divided as follows. In section 2, we illustrate the problem by providing several scenarios. Section 3 will provide a brief technical background. Section 4 will show similar related works. In section 5, we will present our proposed solution.

2 Scenarios

In this section, we present two scenarios. The first scenario presents the behavior of a PHR system during an emergency. The second scenario shows how an adversary might be able to abuse such a system to invade a user's privacy.

In the first scenario, a hospital gets a call about a car accident. They are told that the patient is unconscious, and that he has been severely injured. The hospital rapidly sends an emergency team to the location of the accident and retrieve the patients file before he gets to the emergency. An employee at the hospital enters the patient's ID into the system and then the system would retrieve the patient's EMRs in a reasonable amount of time.

In the second scenario, an employee at a hospital with malicious intentions retrieves a public figure's EMRs. The purpose of this request is to smear that person's reputation by convincing the public he is not healthy enough for a higher position. The employee would enter the patient's ID into the system. Rather than getting the medical records back, the system should reject that request and flag the hospital for abusing the system.

3   Technical Background

The use of a username and password method of authentication would not be useful since we are assuming the PHRs are stored on an untrusted cloud. Since we are assuming the cloud to be untrusted, we assume that PHRs are available to any entity on the cloud. However, the PHRs would be encrypted with ciphertext-policy attribute based encryption. In this section we will provide a more detailed description of attribute based encryption and threshold encryption.

3.1   Attribute Based Encryption

When using Biometrics as identities for Identity Based Encryption, the values of the biometric feature readings contain noise. Amit Sahai and Brent Waters [1] tackled this problem and in doing so, they also introduced what is now known as Attribute Based Encryption. They suggested that if we use w as the Biometric Identity, then the next reading of the biometric features would be w', which would be close to the value of w. They also proposed a Fuzzy Identity Based Encryption scheme which can use w' as an Identity to decrypt messages encrypted using w as the Identity. They then explain that Fuzzy-Based Encryption can be used as an application for "Attribute-Based Encryption". In ABE, ciphertexts are associated with attributes. To Decrypt the ciphertexts, a user must have all the attributes needed to decrypt the file.

Matthew Pirretti, Patrick Traynor, and Patrick McDaniel [2] later proposed a more secure Attribute Based Encryption method. They implemented more complex policies for Attribute Based Encryption based on the work done by Sahai et al. [1]. In their method they allow complex policies such as having 'and' and 'or' logical policies. They also applied their method to a medical application where the patients' medical records were only available to entities with the proper attributes.

There are two basic ABE schemes, namely cipher-based attribute based encryption (CP-ABE) [3] and key-policy attribute based encryption (KP-ABE)[4]. The difference between these two schemes is what the attributes describe. In CP-ABE the attribute access structures are used to describe the encrypted data, while in KP-ABE the attribute access structures are used to describe the user's key. CP-ABE is similar to RBAC, while KP-ABE is similar to ABAC.

3.2   Threshold Cryptosystems

In a (t, n) threshold scheme [5], a secret S is split into n shares and distributed to the participants. To reveal the secret S, any t participants must work together and use their share to calculate the secret S. Threshold schemes also ensure that when $t - 1$ participants work together, it would not be possible for them to calculate S nor gain any information about it.

4   Related Work

Ming Li et al [6] proposed a method for providing access to emergency PHRs encrypted with ABE during an emergency. They suggested using a break-glass approach where each patient's PHR's access rights are also delegated to an emergency department. To protect the PHR break-glass option from being abused they suggested that the ER staff need to contact the emergency department to verify their identities and verify the emergency case. The emergency department would then give temporary read keys. After the emergency, the patient may restore normal access by computing another key and then submit it to the emergency department and the server to update the files.

Although their proposed solution provides access in emergency cases, they do make the assumption of trusting emergency departments. Any employee in an emergency department would be able to access any emergency PHR record in the system. In our work we do make the assumption that non of the entities in the system are completely trusted. We also make the assumption that the access to PHRs is not achieved by emergency staff accessing them using their personal keys. In our work we simplify the process by giving all ECPs the same attributes and keys.

Gardner et al. [7] proposed a different method which did not rely on attribute based encryption. In their work they assumed the PHRs are to be stored on a smart phone, and their goal was to protect the PHRs stored on the phone from being accessed by an adversary. Moreover, the approach they proposed may be modified to access EMRs stored on an untrusted cloud. They proposed using Secret Sharing [5] to split the decryption key and distribute it to different entities. By dividing the access capabilities, it becomes possible to associate the credentials of different entities with varying weights. The partial weighted rights are then distributed in a way where no one credential is sufficient to obtain access, but appropriate combinations are.

They used Secret Sharing to distribute the decryption key k_r between the following entities:

- Break-The-Glass (BTG): a share that is used only as a last-resort access mechanism.
- Password: a share accessible with the PHR owner's password.
- EP: a share that is available to ECPs.
- Face: a share that is accessible with the PHR owner's face.
- Finger: a share that is accessible with the PHR owner's fingerprint.

Although this solution could be applied to provide access to PHRs stored on an untrusted cloud. We do not think it could be used as a global solution because this would mean that all medical health providers would be using the same key. This implies that any ECP would be able to view all PHRs on the cloud. Gardner et al. discussed this problem and suggested three mehtods for solving this problem.

- The first was to provide every EP with a different key pair. However, this would mean that the client software must store a corresponding ciphertext for each key. This would obviously have a huge storage overhead if looked at as a global solution.
- The second solution was to use a public-key broadcast encryption system such as that proposed by Boneh et al.[8], a trusted entity would generate a private key for every EP and a broadcast public key. This entity would program the private keys onto a smart card for each EP and maintain a list of non-revoked keys. The client software would use this list and the broadcast public key to encrypt the EP share.
- The third solution was to update the EP keys everyday, but this would also require the patient to download the new keys and re-encrypt the PHRs everyday.

Narayan et al [9] proposed a scheme for Privacy preserving electronic health records (EHR) using an attribute-based infrastructure. The system they proposed is patient-centric which means that the patient has complete control over the EHR. In their scheme the EHR are stored in untrusted cloud storage. The EHR in this system are seen as a directory of files stored in sub-folders. They assumed EHR to have the following structure:

- Patients' health data in the form of encrypted files. These files are encrypted with symmetric keys which can be found in the entry files.
- A table consisting of entries corresponding to the patient's files. The contents of the entry files include the following:
  - Meta data describing the files and their locations, all in encrypted form. The data is encrypted using broadcast ciphertext-policy attribute-based encryption and includes the following information:
    * File description.
    * A random locator tag which is also the file name used to locate the file in the cloud.
    * A symmetric key used to encrypt the health data.
  - An access policy in plain text form which specifies who decrypt the encrypted data in the entry file.
  - A search-index for keywords within the encrypted file used for keyword search. The search index is the encryption of the keyword.

5   Proposed Solution

We base our work on the privacy preserving PHR scheme proposed by Narayan et al. [9] described in the related work section. However, in our work we introduce a method that would allow ECPs to get access to a patient's EMRs in case of an emergency without giving them access to all EMRs in the system.

In our scheme, rather than generating a single key for every user in the system, we propose generating another ABE key pair to aid in the emergency access protocol. The public key generator (PKG) would generate another key k_{ep} for

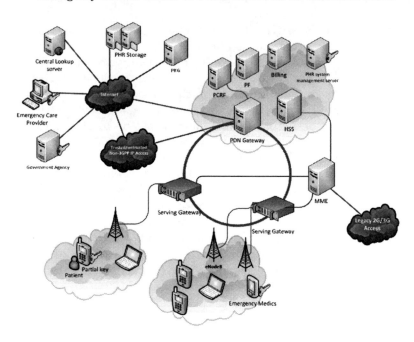

Fig. 1. Proposed system architecture under an LTE mobile network

every patient in the system for use in case of an emergency. This key would be generated under the policy:

$$\{emergency \wedge patient'sID = x\}.$$

In our proposed scheme, we encrypt all EMRs with k_{ep}. We also note that no one in the system would have that key, not even the ECPs or the patient. The patient would not be able to decrypt the records without having the emergency attribute.

The PKG would then split the k_{ep} key into n shares, where k shares would be needed to recalculate k_{ep}. To achieve this, we propose using shamir's (k,n) threshold scheme [5]. In our scheme, we propose splitting the keys into four shares, where any two shares are needed to compute k_{ep}. However, increasing the number of shares needed to access an EMR would make it more difficult for an attacker to conspire with key share holders to unlawfully access a patient's EMRs. All the key shares would also be encrypted with k_e, which is a key with the policy $\{emergency\}$. This would make the key shares accessible to ECPs only. The key shares will be given to the following entities:

- One share would be in the patient's entry file. This would enable any ECP to access the patient's first k_{ep} key share. However, the ECPs would not be able to decrypt the EMRs because they would still need another share of k_{ep} to be able to decrypt the EMRs.

- A second share of k_{ep} is given to the patient for cases where the patient needs emergency medical care, but is also able to participate in the protocol. In this case, the protocol would enable the patient to provide the ECP with the second share enabling them to retrieve and decrypt the emergency medical records. This share could be stored on the patient's mobile device and sent to the ECP after the patient acknowledges the emergency.

  Another method of storing the key share could be to give the patient an RFID chip with the key share stored on it. This could even save time if found in the patient's wallet during an emergency.
- The third share would be giving to a third party, which would provide the third share after verifying that the patient needs emergency medical attention. In this scheme, we assume the third party to be the patient's telecom provider. We do not assume that the telecom provider is completely trusted, which is why all the key shares are also encrypted with k_e. In section $4.2.5$ we will further discuss why we chose the telecom providers to be the third party and the methods they may use to verify an emergency situation. After verifying the patient's need for emergency medical attention, the partial key is sent to the ECP allowing them to calculate k_{ep} and decrypt the emergency medical records.
- The fourth share would be given to a government agency which would provide the fourth share after verifying that the patient is in need of emergency medical care. This verification would be done manually by a government employee. After making sure that the patient is in need of emergency medical attention, the government agency would authorize the system request to send the k_{ep} key share to the ECP.

Encrypting the EMRs with an ABE key as in our proposed solution would enable the patient to modify the medical records at anytime. This is mainly because the system would automatically generate the EMRs from the patient's medical records and encrypt the EMRs with the policy set to:

$$\{emergency \wedge patient'sID = x\}.$$

Although encrypting the EMRs with an ABE key gives us the advantages we mentioned above, we find that a problem could arise after a patient has received the needed emergency care. Since the same key would always be used to encrypt the patient's EMR, the ECP would always be able to access the patient's medical records. If we assume that this could be a security risk, we propose the following solutions to solve this problem:

- Session attributes could be added to the ABE access policy. The new policy would be:

$$\{emergency \wedge patient'sID = x \wedge session = y\}.$$

It would be enough for the session attributes to be a counter where every time the patient gets emergency treatment the counter y is increased by one. The PKG would also need to split the new key to four share and redistribute them.

- Rather than using an ABE key k_{ep} to encrypt the EMRs, a random symmetric key Kr would be used to encrypt the patients' EMRs. Similar to our protocol, the first share of Kr would also be encrypted with k_e and placed in the patient's entry file. The other shares would be distributed to the other parties after encrypting them with k_e.

Although using a symmetric key might seem more appropriate than using an ABE key as in our proposed solution, we find that when using the latter, the shares could always be recomputed by the PKG in case one of the key shares is lost. However, if one of the symmetric key shares is lost then nothing can be done unless a copy of the entire symmetric key is stored somewhere.

5.1 Key Share Providing Protocol

We propose an authentication protocol in place for sending a key share to the ECP. The purpose of this protocol is to authenticate both parties to each other, and ensure that the key share is securely sent to the ECP. ABE is used in this protocol to ensure the confidentiality and integrity of the data.

Notations:

- ECP: Emergency Care Provider
- KSH: Key Share Holder (this could be any entity with a key share, ex. patient or telecom provider)
- PID: Patient's ID
- $ECPID$: ECP's ID
- $KSHID$: KSH's ID
- $\{..\}_{K_{KSH}}$: Encrypted with a key which has the identity attributes set to the KSH's ID
- $\{..\}_{K_{ECP}}$: Encrypted with a key which has the identity attributes set to the ECP's ID
- K_{epi}: A key share of K_{ep}

The Key Share Request Protocol:

- $ECP \rightarrow KSH$ $\{PID, ECPID\, RAND1\}_{K_{KSH}}$
- $KSH \rightarrow ECP$ $\{PID, RAND1, RAND2\}_{K_{ECP}}$
- $ECP \rightarrow KSH$ $\{PID, RAND2\}_{K_{KSH}}$
- $KSH \rightarrow ECP$ $\{PID, \{K_{epi}\}\}_{K_{ECP}}$

5.2 Building a Patient-Centric EHR System

In this section, we outline the algorithms needed to construct a PHR system with the properties described in our proposed solution. The protocol is presented in a sequence diagram in figure 2.

The main concept of this work is to build on current ABE PHR systems to allow emergency access. Access to the EMRs is to be provided only in case of a verified emergency. Granting access should not allow ECPs the right to view every emergency medical record in the system. The encryption scheme of our PHR system is based on CP-ABE [3] and contains the following functions:

PHR-Setup: Runs CP-ABE-Setup(): It is used to obtain the system public key PK and the master secret MK. The setup algorithm will choose a bilinear group $\mathbb{G}_0$ of prime order p with generator g. Next it will choose two random exponents $\alpha, \beta \in \mathbb{Z}_p$. The public key is published as:

$$PK = \mathbb{G}0, g, h = g^\beta, e(g, g)^\alpha$$

and the master key MK is (β, g^α).

PHR-KG(S,ID): Run CP-ABE-KeyGen(MK, S, ID). The key generation algorithm will take as input a set of attributes S along with the ID of the user, and outputs a secret key SK_{id} that identifies with the set of user attributes.

PHR-ER-KG(ID): Run CP-ABE-KeyGen($MK, emergency, ID$) to generate an emergency key k_{ep} for a patient, the key generator will generate a key k_{ep} with the attributes $\{emergency, patientID\}$.

PHR-Split-ER(k_{ep}): It is used to split the key k_{ep} into four shares where any two shares are enough to reconstruct the key k_{ep}. The algorithm first chooses a random value $rand$. Then it forms the polynomial:

$$f(x) = k_{ep} + rand\, x$$

Finally, it computes and returns the key shares $k_{epi} = f(i)$, $1 \leq i \leq 4$, along with the index i.

PHR-Combine-ER(k_{ep1}, k_{ep2}) it combines any two shares to reconstruct k_{ep}. The algorithm computes the coefficient a_1 of f(x) by using Langrangian interpolation. The algorithm then can compute $f(0) = a_0 = k_{ep}$.

PHR-Encrypt: Run CP-ABE-Encrypt($PK, \mathcal{M}, \mathcal{T}$) to generate a cipher text of a message $\mathcal{M}$ under the tree access structure $\mathcal{T}$.

PHR-Decrypt: Run CP-ABE-Decrypt(CT, SK_{id}). The decrypt algorithm takes as input the ciphertext CT and the private key SK_{id}.

5.3   Emergency Verification

In our proposed solution, we suggested that a third party would provide a key share of k_{ep}, which is also encrypted with k_e, to the ECP after verifying that the patient is in need of medical attention. It is difficult to come up with an

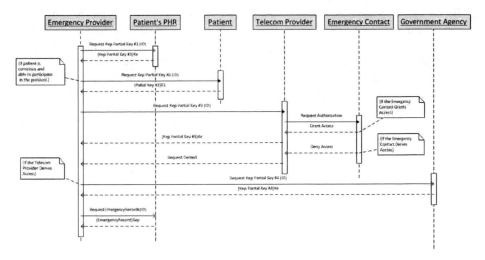

Fig. 2. Sequence diagram of proposed solution

automated method to achieve this task because a false negative could delay the medical care a patient needs. Also, a false positive would be an invasion of a patient's privacy. There are numerous ways of verifying an emergency. We propose in this section three possible methods of emergency verification.

Willkomm et al. [10] analyzed data collected over three weeks at hundreds of base stations in highly populated areas. The dataset consisted tens of millions of calls and billions of minutes of talk time. In their study they found that 10% of RF voice channels were allocated for the duration of about 27 seconds. They later confirmed that these were calls that were either not answered and then redirected to voice mail. This means that 90% of all phone calls were answered. With a 90% probability of having phone calls answered, we propose giving the third share to a third party that would contact the patient and emergency contacts to verify the emergency. The third party could be a telecom provider given the resources they have. Moreover, with the high percentage of mobile subscribers, we are assuming every user in the system would be a mobile subscriber. We propose the following three methods of verification, each with its advantages and disadvantages.

Location Based Verification: Telecom providers have the ability to find a mobile subscribers approximate location given the cell towers providing the mobile subscriber with a connection to the network.

When a ECP requests a key share for a patient, the telecom provider may provide it if the patient's telecom device is in a location within a reasonable distance from the hospital's location.

This solution has the benefit of being completely automated. This would lead to a response within seconds, allowing the ECPs the chance to act rapidly to save the patient's life.

Although this solution would deny access from ECPs which are not in the same location as the patient's mobile device, it would still be possible for an

ECP to target a specific person within the same location where the patient's mobile device is located. Access might also be denied if the patient's mobile device's battery runs out in a location far from the location of the emergency. This would mean that the telecom provider's data shows the last known location of the patient's device in a different area. However, that could also be solved with an algorithm that would estimate the possibility a patient could travel from the last known location to the location of the emergency. Also, access to a patient's key share might be denied because the patient did not have his/her mobile device at the location where the emergency occurred.

Emergency Contact Verification: To overcome the disadvantages in the location based solution, we propose contacting the patient's emergency contacts. The telecom provider could either send a text message or generate an automated call to the patient asking for permission to provide access to the ECP. In case the patient does not reply within a reasonable amount of time, the same request would be sent to the patient's emergency contacts.

This method has the benefit of being more thorough in verifying whether the patient is in need of emergency care or not. However, it does require more time, which could be dangerous to the patient's health in an emergency. Moreover, a request might be denied if the emergency contacts are contacted during late hours of the night.

Location-Based and Emergency Contact Verification: Both of the emergency verification methods we mentioned above had their advantages and disadvantages. Thus, we propose using a hybrid solution that combines both methods. We propose using the emergency contact verification method in the beginning, and in case no response is returned, the location based method would take over. This means that a request would only be denied if the emergency contacts do not respond within a reasonable time interval and the patient's last location is not in proximity to the ECPs location.

6   Conclusion

In this work we discussed the problem of providing emergency response providers access to emergency PHRs stored on an untrusted cloud. We proposed a solution based on using ABE and threshold cryptography to secure the data on the cloud while providing fine grained access control and confidentiality. In our solution we proposed a protocol capable of providing access to these records, but only during a verified emergency. We achieved this by splitting the decryption key between four parties where two are needed to calculate the decryption key. We also discussed how verifying the occurrence of an emergency can be automated by a third party such as a telecom provider. We believe that this method can be generalized to storing sensitive data in untrusted storage systems such as cloud storage systems.

References

1. Sahai, A., Waters, B.: Fuzzy identity-based encryption. In: Cramer, R. (ed.) EUROCRYPT 2005. LNCS, vol. 3494, pp. 457–473. Springer, Heidelberg (2005)
2. Pirretti, M., Traynor, P., McDaniel, P., Waters, B.: Secure attribute-based systems. In: Proceedings of the 13th ACM Conference on Computer and Communications Security, CCS 2006, pp. 99–112. ACM, New York (2006), http://doi.acm.org/10.1145/1180405.1180419
3. Bethencourt, J., Sahai, A., Waters, B.: Ciphertext-policy attribute-based encryption. In: Proceedings of the 2007 IEEE Symposium on Security and Privacy, SP 2007, pp. 321–334. IEEE Computer Society, DC (2007), http://dx.doi.org/10.1109/SP.2007.11
4. Goyal, V., Pandey, O., Sahai, A., Waters, B.: Attribute-based encryption for fine-grained access control of encrypted data. In: Proceedings of the 13th ACM Conference on Computer and Communications Security, CCS 2006, pp. 89–98. ACM, New York (2006), http://doi.acm.org/10.1145/1180405.1180418
5. Shamir, A.: How to share a secret. Commun. ACM 22(11), 612–613 (1979), http://doi.acm.org/10.1145/359168.359176
6. Li, M., Yu, S., Ren, K., Lou, W.: Securing personal health records in cloud computing: Patient-centric and fine-grained data access control in multi-owner settings. In: Jajodia, S., Zhou, J. (eds.) SecureComm 2010. LNICST, vol. 50, pp. 89–106. Springer, Heidelberg (2010)
7. Gardner, R.W., Garera, S., Pagano, M.W., Green, M., Rubin, A.D.: Securing medical records on smart phones. In: Proceedings of the First ACM Workshop on Security and Privacy in Medical and Home-Care Systems, SPIMACS 2009, pp. 31–40. ACM, New York (2009), http://doi.acm.org/10.1145/1655084.1655090
8. Boneh, D., Gentry, C., Waters, B.: Collusion resistant broadcast encryption with short ciphertexts and private keys. In: Shoup, V. (ed.) CRYPTO 2005. LNCS, vol. 3621, pp. 258–275. Springer, Heidelberg (2005)
9. Narayan, S., Gagné, M., Safavi-Naini, R.: Privacy preserving ehr system using attribute-based infrastructure. In: Proceedings of the 2010 ACM Workshop on Cloud Computing Security Workshop, CCSW 2010, pp. 47–52. ACM, New York (2010), http://doi.acm.org/10.1145/1866835.1866845
10. Willkomm, D., Machiraju, S., Bolot, J., Wolisz, A.: Primary users in cellular networks: A large-scale measurement study. In: New Frontiers in Dynamic Spectrum Access Networks, DySPAN, pp. 1–11 (2008)

Generating Evidence-Based Pathway Models
for Different Hospital Information Systems

Katja Heiden and Britta Böckmann

University of Applied Sciences and Arts Dortmund, Emil-Figge-Straße 42,
44227 Dortmund, Germany
{katja.heiden,britta.boeckmann}@fh-dortmund.de

Abstract. Healthcare providers are facing an enormous cost pressure and a scarcity of resources, so that they need to realign in the tension between economic efficiency and demand-oriented healthcare. Clinical guidelines and clinical pathways have been established in order to improve the quality of care and to reduce costs at the same time. Clinical guidelines provide evident medical knowledge for diagnostic and therapeutic issues, while clinical pathways are a road map of patient management. The consideration of clinical guidelines during pathway development is highly recommended. But the transfer of evident knowledge (clinical guidelines) to care processes (clinical pathways) is not straightforward due to different information contents and semantical constructs. This article proposes a model-driven approach to support the development of guideline-compliant pathways and focuses the generation of ready-to-use pathway models for different hospital information systems. That way, best practice advices provided by clinical guidelines can be provided at the point of care and therefore improve patient care.

Keywords: Clinical Pathways, Clinical Guidelines, Metamodeling, Health Level 7, Ontologies, Hospital Information Systems.

1   Introduction

Clinical guidelines and clinical pathways are accepted instruments for the quality assurance and process optimization in the healthcare domain. Both concepts define a standardized best practice about appropriate patient treatment for a specific disease. Clinical pathways are defined as "a complex intervention for the mutual decision making and organization of care processes for a well-defined group of patients during a well-defined period" [22]. They provide a process-like description of proper medical treatment, whereas clinical guidelines are defined as "systematically developed statements to assist practitioner and patient decisions about appropriate health care for specific clinical circumstances" [5]. They represent results of latest research. The positive impact of clinical guidelines on the quality of care has been scientifically proven in [6]. But their influence on clinical routine is still very low in Germany due to their narrative and non-formalized form [18]. A decisive factor for the success of clinical guidelines is the provision of the knowledge at the point of care [13].

G. Huang et al. (Eds.): HIS 2013, LNCS 7798, pp. 42–52, 2013.
© Springer-Verlag Berlin Heidelberg 2013

The main challenge can be summarized as follows; imprecise, not formalized and abstract guidelines have to be implemented as concrete processes. Clinical pathways are appropriate for that purpose; they are used in different kinds of healthcare facilities, facing a diversity of hospital information systems (HIS). The following approach presents an IT system, which allows the derivation of guideline-compliant pathways and which generates evidence-based pathway models for different target systems.

2   Conceptual Model Development

The consideration of clinical guidelines during pathway development is highly recommended, but there are no standardized mechanisms, which ensure a guideline-compliant care. Guideline recommendations are abstract and therefore not directly applicable. Thus, these recommendations have to be implemented and tailored to local settings. By using clinical pathways for that purpose, the content-adaption of clinical guidelines leads to concrete process flows [13]. That way, the latest scientific findings can be applied in everyday healthcare and therefore lead to a best quality of patient care. This section presents the conceptual model to ensure the development of guideline-compliant pathways.

2.1   Related Work

Several methodological approaches exist to implement clinical guidelines into operational practice. They show considerable differences concerning the aim or result of the translation process (clinical pathways vs. computer-interpretable guidelines). In addition, they vary in the degree of automation (highly manual vs. semi-automated approaches).

One approach focus the formalization of narrative guidelines in a computer-interpretable form (see [11, 21]), which can be processed in decision support systems. The domain experts are supported during this process by special editors. It is error-prone to map prose text to coded data, because clinical guidelines can partly be ambiguous, incomplete, or even inconsistent [14]. Several guideline representation languages exist as rule-based languages, e.g. Arden Syntax, logic-based languages, e.g. PROforma, task network models, e.g. Asbru, GLIF, or document-centric approaches i.e. GEM. Differences and similarities of these languages are outlined in [16, 19]. If computer-interpretable guidelines should be used by a hospital in order to provide the medical knowledge during patient care, the hospital information systems need to have the ability to interpret and use those formalizations. The translation process does not produce a clinical pathway by definition; rather computer-interpretable guidelines (CIGs) are created to support the decision making process during patient treatment. This approach provides one way to implement guideline recommendations in daily routine, but, according to [13], the translation of clinical guidelines into alerts and reminders does not support the patient treatment as a unit.

The second approach is a highly manual process, where clinical pathways are developed on the basis of existing clinical guidelines (see [2, 8, 15]). The pathway development process is done by a group of domain experts. An interdisciplinary team is composed of all professional groups involved, e.g. physicians, nurses, medical controllers, quality assurance representatives. At this approach, the pathway development process starts with an extensive literature research, where pertinent guidelines are identified. The analysis needs to be done manually by the interdisciplinary team or a single project member. It is a time-consuming and resource-intensive task with no methodical support. The extracted recommendations from the guidelines can be used as an input for the pathway development. Therefore, the guideline content needs to be tailored to local conditions and a consensus among the participating health professionals needs to be reached [13]. Information technology is mainly used for modeling tasks and not for the whole lifecycle management of clinical pathways. The result of this development process is a clinical pathway for one specific healthcare facility. The interdisciplinary team produces text documents or informal process models, which describe the appropriate care for a specific disease in form of a clinical pathway. Thus, the developed pathways cannot be directly interpreted by IT systems. A formalized description and additional technical information for the enactment of clinical pathways needs to be defined, e.g., mapping of service calls to specific tasks. The implementation of clinical pathways is a separate step, which is done by IT-specialists. It is an error-prone task, because these experts often do not have detailed domain knowledge and sometimes the pathway definitions are ambiguous. There is a high need for communication between the domain- and the IT-specialists. Several cycles are necessary to implement the pathways in the present information system. Thus, a gap between development and implementation of clinical pathways exists as well as a media break between both process steps.

A further approach focuses the systematic derivation of clinical pathways from clinical guidelines by the help of a model-based methodology (see [10, 20]). Jacobs [10] developed a reference model for the methodical transfer of clinical guidelines in clinical pathways, which was exemplary deduced from the breast cancer treatment. One universal pathway for a specific guideline was derived, which can be adapted to a special institution in a further step. Jacobs performed a theoretical examination of the derivation process; but no consistent IT support for the implementation of the derived clinical pathways in the present information system was provided. Schlieter [20] carried the results from Jacobs forward. Rule sets were added to the reference model in order to define the way of reusing model content. Schlieter provides construction techniques for deriving specific models from a formalized guideline.

After presenting the related approaches, the following requirements can be imposed, as a quintessence, for our work. In contrast to other approaches, the whole development process (definition, implementation, lifecycle management) is supported. Each presented approach covers only one aspect of the entire derivation process; a formal representation (see CIGs), the development of concrete clinical pathways for one specific institution (manual process), or the systematic translation (model-based approach). The work in hand creates a guideline-compliant clinical pathway, which describe the whole patient treatment of a special disease in one concrete setting. Additionally, ready-to-use path models are generated for different hospital information systems to implement the derived pathways. Related approaches especially neglect the last step (deployment), although it is a decisive factor for the success and usage of clinical pathways.

2.2   Model-Driven Approach

We propose a model-driven approach for defining guideline-compliant pathways (see Fig. 1). The key part is an underlying metamodel, which provides a formalized description of evidence-based pathways and stores all extracted information. A metamodel offers the ability to represent data structures of a special domain in a concise way. In this context, it can merge knowledge of clinical guidelines and clinical pathways in a generic model before the guideline-compliant pathways are integrated in concrete information systems. Our goal is the provision of a metamodel that underlies both; the knowledge elicitation process (step 1 in Fig. 1) and the generation process (step 2 in Fig. 1). A detailed description of the derivation process can be found in Sec. 2.3. The entire model-driven approach for defining guideline-compliant pathways is supported by information technology. Therefore it consists, in general, of two different modules; a modeling component to define a clinical pathway on the basis of existing clinical guidelines and a generation component, which supports the deployment of these pathways in concrete information systems.

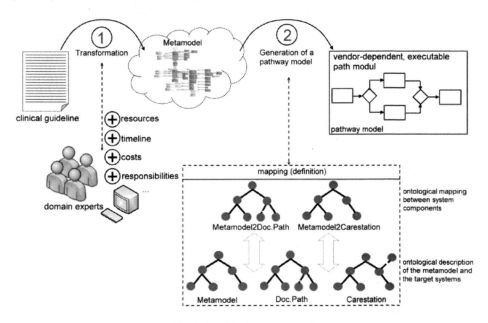

Fig. 1. Model-driven approach

The encoding for the metamodel is given through artifacts of Health Level 7 (HL7), a worldwide accepted standard in the healthcare domain. HL7 provides a comprehensive framework for interoperability that improves care delivery, optimizes workflow, reduces ambiguity, and enhances knowledge transfer among all of the stakeholders; see [9]. It ensures a platform independent description of guideline-compliant pathways until they are deployed in concrete information systems. The encoding was done on basis of the HL7 Care Plan Model, which has been developed by the HL7 group to define action plans for various clinical pictures. The HL7 Care

Plan Model is not a normative standard yet [9]; rather it is a draft, which can be extended and adapted to depict the metamodel of guideline-compliant pathways.

Before we point out, how the derivation of guideline-compliant pathways takes place, we briefly present the metamodel. It can describe intersectoral pathways involving multiple medical disciplines and healthcare facilities. Clinical guidelines provide, in the majority of cases, recommendations for the whole medical treatment of a disease including different episodes of care, e.g. prevention, diagnostics, therapeutics, follow-up, and rehabilitation. Our model offers elements to depict such an integrated care approach. Every episode of care can be supported by different healthcare facilities and is represented by a set of clinical pathways. These pathways describe the entire care activities for one episode. Fig. 2 exemplary summarizes that; the episode *acute care* is completely described through the pathways P3 – P7. Each pathway can furthermore be divided into different treatment phases, e.g. a preoperative day, a day of surgery, and a postoperative day, see Fig. 2. Each phase is described by a set of medical, nursing, and administrative activities. According to Health Level 7, there exist different activity types to describe the patient treatment, e.g. procedures, medications, encounters, and observations. A pathway can be defined by composing these activity types, adding responsibilities in the care process, and specifying the detailed control flow. To realize the last-named routing, we integrated the HL7 workflow control suite of attributes. Beside these process-like structures, additional information from clinical guidelines, e.g. risk factors, complications, or guiding symptoms, is considered. This information has no influence on the control flow of patient treatment, but it can create added values especially for young professionals and thus should be displayed on demand in the target systems. To depict this information, a generic parameter system has been integrated in the metamodel.

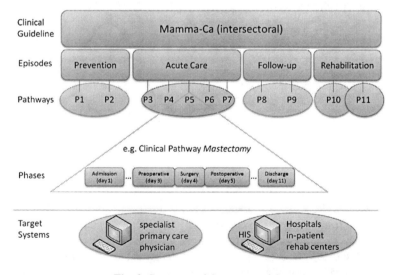

Fig. 2. Structure of the metamodel

The HL7 Care Plan Model was extended by the following classes to depict the entire elements of the metamodel:

1. Assignment of clinical guidelines to clinical pathway; that way the evident basis for the definition of a guideline-compliant pathway is stored
2. Adding structural components to define intersectoral pathways, and to divide the whole patient treatment in coherent units, e.g. episodes of care
3. Defining concepts like costs, strength of evidence and recommendation, and so on (those contents cannot be depicted by the original model)
4. Integrating the HL7 Control Suite of Attributes to enable the definition of a detailed control flow
5. Representing additional information within a parameter system

These extensions should be evaluated and proved in a HL7 interoperability forum, which intends the presentation and discussion of HL7 specific implementations.

2.3   Derivation Process

The derivation process is done by an interdisciplinary team. They use existing (narrative) clinical guidelines to define guideline-compliant pathways. In general, three different steps need to be performed by the domain experts.

- Step 1: Extraction of all pertinent guideline recommendations, which should be considered during pathway development, and which are classified based on the elements of the metamodel. The classification indicates the content of a narrative recommendation, e.g. medication, procedure, observation, additional information. We deliberately avoid the direct mapping of these narrative recommendations onto coded data of the metamodel. The coding is done in two different steps. Step 1 only includes the extraction of pertinent information from a clinical guideline, which is supported by an integrated mark-up tool. In addition, we enriched the narrative recommendation semantically by classifying the content. In step 2 the coding is realized.
  Example: *"Patients with early invasive breast cancer should have a baseline dual energy X-ray absorptiometry (DEXA) scan to assess bone mineral density"* [17].
  Classification: An X-ray is an examination (observation).
- Step 2: Formalization of the recommendations; through the classification in the previous step, the correlation between the metamodel and the guideline excerpts is already given. The metamodel defines all descriptive attributes for a specific element, e.g. a procedure consists of a name, an OPS code, and so on. Based on this definition, customized forms are generated to capture the remaining information in order to formalize the narrative recommendations. Step 2 is completed by adding further activities, which are not defined within clinical guidelines, but which need to be performed during patient treatment, i.e. nursing activities. After step 2 all activities are represented in a computerinterpretable form (HL7).
  The following representation of the DEXA examination is produced by the IT system (see Fig. 3). The domain experts do not need any knowledge about the metamodel and its representation. They use generated forms to fill in all necessary data and the system stores the formalized elements within the structures of the metamodel.

```
<ObservationDefinition classCode="OBS" moodCode="DEF">
   <title>dual energy X-ray absorptiometry</title>
   <code codeSystem="2.16.840.1.113883.6.11" code="38268-9" codeSystemName="LOINC" />
</ObservationDefinition>
```

Fig. 3. HL7 representation of a dual energy X-ray absorptiometry

- Step 3: Composition of a guideline-compliant pathway by sequencing the defined activities, by adding responsibilities and resources, by defining a detailed control flow and so on. This step is supported by a graphical editor. The input for step 3 is the list of all formalized activities (step 2). Based on these elements, the definition of the clinical pathway can be done.

The IT system supports the domain experts in extracting pertinent knowledge from clinical guidelines and mapping the information onto the underlying metamodel. The main goal is to hide the complexity of the metamodel and provide an intelligent guidance to the domain experts to extract pertinent elements from the narrative guidelines and to translate these recommendations in pathway structures. The domain experts need to have the ability to use the IT system without having knowledge about the underlying model and the target representation of the backend system. Therefore the acquisition and definition process can be performed without imposing the burden on the experts of learning formal representation languages. A clinical guideline is one knowledge source during this process and is enriched by factual and practical knowledge of the domain experts, which has a bearing on the development process as well.

3   Generation of Ready-to-Use Pathway Models for Different Hospital Information Systems

After the development process, the guideline-compliant pathways are defined and represented within the metamodel. Most of the healthcare facilities use and manage clinical pathways in their information systems. Therefore the next step is the deployment of the guideline-compliant pathways. The introduction of clinical pathways into a hospital information system can create the most added values, i.e. a continuous surveillance of clinical outcome is possible or routinely documentation tasks can be reduced, because only derivations from a pathway need to be recorded. There exists a diversity of HIS in German hospitals. They use different strategies to manage and enact clinical pathways. Some HIS use explicit process models in conjunction with a diversity of process modeling languages, e.g. Business Process Model and Notation (BPMN), or XML Process Definition Language (XPDL). Other store clinical pathways within the HIS database. The guideline-compliant pathways, which are depicted within the metamodel, are described by a HL7 specific XML structure. As a consequence, to execute the guideline-compliant pathways, they need to be translated into different formalisms of the target systems, i.e. SQL or various XML syntaxes. For that purpose, two steps are required; a mapping between the elements of the metamodel and the elements of the target systems needs to be done; see Sec. 3.1. In addition, a technical translation of the pathway into the target formalism has to be realized; see Sec. 3.2. Fig. 4 points out different aspects of the generation process.

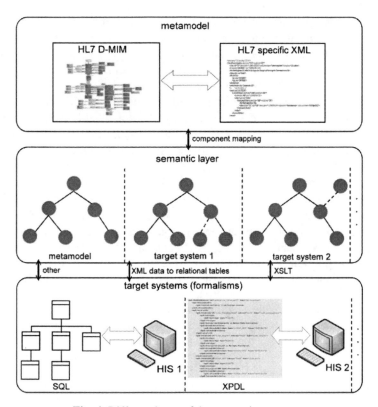

Fig. 4. Different layer of the generation process

3.1 Component Mapping

We take a knowledge management approach to ontologically model the component mapping between the elements of the metamodel and the elements of the target systems. Gruber [7] defines an ontology as an "explicit formal specification of a shared conceptualisation". Ontologies represent a key technology, which enable semantic interoperability and integration of data and process [3]. "Semantic" means the mapping between a language or modeling syntax to some formalism, which expresses the "meaning" [1]. Semantic applications do not focus on the presentation of data rather they focus on the subjects and their relationships. The main advantage is the explicit representation of these concepts, that underlie an application and the needed presentations are generated by these specifications [1]. In this approach, different metadata schemes of HIS describing clinical pathways need to be linked. The main advantage of modeling the component mapping ontologically instead of hard coding this information content is the possibility to extend or adapt the knowledge consecutively. If the data model of a target HIS changes, these modifications can be made intuitively. For performing the initial mapping, a detailed knowledge about the target systems is required. Therefore, a system analysis is performed for every target HIS identifying the available components to represent the information contained in the metamodel.

The component ontology ensures that every element of the metamodel is mapped to one or more equivalent representations in a target system. It is possible that an element of the metamodel has only one counterpart, e.g. a medication. But it is possible too, that there are several ways to represent an element in a target system. This will be clarified by the following example.

The HIS iMedOne (Tieto) provides the path module Doc.Path to manage and enact clinical pathways. The elements of the pathway are scheduled in a calendric view and are assigned to different dimensions. A dimension aggregates equal element types, e.g. orders, medications, nursing activity, so that various views can be defined for different user groups. The element *medication* in the metamodel has only one equivalent in Doc.Path; a prescription. But additional information out of the generic parameter system could be represented in different ways, e.g. a new dimension *information* can be created and the content can be placed within such an element, or this information is displayed within the textual description of a specific path element, or it is represented by a document or an URL, which can be added to every path element.

Based on the possible representation forms defined within the component mapping, the interdisciplinary path team has the choice, how to implement the guideline-compliant pathway. Again the users are supported by the IT system, because the system knows about the available facilities of the target system and can therefore offer the possible elements to the domain experts, who can perform the configuration of the pathway. As a consequence, the arrangement of the pathway models in a special system is no black box for the domain experts. In contrast, they can influence the results by interacting with the IT system. The arrangement of a pathway in a concrete information system may even vary between different users. Thus, the configuration can be stored individually to adapt the generated suggestions to each user.

Concerning the generation of concrete pathway models, there exist, in general, two different kinds of information; active and passive elements. Passive elements, i.e. additional information, can be simply copied to a target system and can be handled as pure data. In contrast, active elements require the creation of special elements in the target system and therefore the interpretation of the information content. For instance, a lab test described within the metamodel, would lead to a concrete order in a target system. Thus, such elements need to be converted and implemented in a backend system.

3.2   Technical Mapping

The last task within our approach is given by a technical translation to actually generate ready-to-use pathways. Basically we have to compile one formalism into another and this kind of problem has been successfully solved over decades through methods of theoretical computer science. So we can use long time approved methods at this point. Coming from the HL7 specific XML structure of the metamodel, there are at least three different translation schemata to mention:

- XML to other XML syntaxes
- XML to a relational data model
- XML to other formalisms

The translation can be realized using a transformational language like XSLT (XSL Transformation), if the target formalism is XML specific either. Additionally, there exist several methods to map XML data into relational tables, cf. [4].

4   Conclusion and Future Work

We presented a model-driven approach, which supports hospitals and other healthcare facilities in developing guideline-compliant pathways. As a result, the latest scientific findings can be transferred into clinical routine. Our goal is to support the entire life-cycle of guideline-compliant pathways by one IT system including the definition and deployment of the pathways in different target systems. The IT system enables the domain experts to model the pathways without requiring knowledge about formal representation languages and to configure the defined pathway in a concrete target system. The encoding of the metamodel by Health Level 7 ensures a non-proprietary solution; the last translation process is not even required for HIS, which can import clinical pathways using a HL7 interface.

The first part of this approach was already proven by one concrete example. We translated the interdisciplinary S3 Guideline for the Diagnosis, Treatment and Follow-up Care of Breast Cancer [12] into the metamodel and showed that the whole information content can be depicted precisely by the HL7 artifacts. In a further step, the IT system should be evaluated in cooperation with a German hospital focusing on the handling of the system. Additionally, the generation process of ready-to-use pathway models should be tested by different HIS.

References

1. Allemang, D., Hendler, J.A.: Semantic Web for the Working Ontologist: Modeling in RDF, RDFS and OWL. Morgan Kaufmann Publishers/Elsevier (2008)
2. Biber, F.C., Schnabel, M., Kopp, I.: Implementierungsstrategien klinischer Pfade. In: Barrierenorientierte Interventionen am Beispiel "proximale Femurfraktur", Marburg (2010)
3. Bloehdorn, S., Haase, P., Huang, Z., Sure, Y., Völker, J., van Harmelen, F., Studer, R.: Ontology Management. In: Davies, J.F., Grobelnik, M., Mladenic, D. (eds.) Semantic Knowledge Management: Integrating Ontology Management, Knowledge Discovery, and Human Language Technologies, pp. 3–20 (2009)
4. Florescu, D., Kossman, D.: Storing and Querying XML Data using an RDBMS. Data Engineering Bulletin 22(3), 27–34 (1999)
5. Grossman, J.H., Field, M.J., Lohr, K.N., Institute of Medicine (U.S.). Committee to Advise the Public Health Service on Clinical Practice Guidelines: Clinical Practice Guidelines - Directions for a New Program. National Academy Press, Washington (1990)
6. Grimshaw, J.M., Thomas, R.E., MacLennan, G., Fraser, C., Ramsay, C.R., Vale, L., Whitty, P., Eccles, M.P., Matowe, L., Shirran, L., Wensing, M., Dijkstra, R., Donaldson, C.: Effectiveness and efficiency of guideline dissemination and implementation strategies. Health Technol. Assess. 8(6), 1–72 (2004)
7. Gruber, T.R.: A Translation Approach to Portable Ontology Specifications. Knowledge Acquisition 5(2), 199–221 (1993)

8. Haeske-Seeberg, H., Zenz-Aulenbacher, W.: Das Sana-Projekt - Geplante Behandlungsablufe. In: Hellmann, W. (ed.) Praxis Klinischer Pfade - Viele Wege führen zum Ziel, pp. 116–179. Ecomed Landsberg (2002)

9. Health Level 7 International (October 08, 2012),
   http://www.hl7.org/about/index.cfm?ref=nav

10. Jacobs, B.: Ableitung von klinischen Pfaden aus evidenzbasierten Leitlinien am Beispiel der Behandlung des Mammakarzinoms der Frau. Dissertation, Medizinische Fakultät der Universität Duisburg-Essen (2006)

11. Kaiser, K., Akkaya, C., Miksch, S.: How can information extraction ease formalizing treatment processes in clinical practice guidelines? A method and its evaluation. Artif. Intell. Med. 39(2), 151–163 (2007)

12. Kreienberg, R.: Interdisziplinäre S3-Leitlinie für die Diagostik, Therapie und Nachsorge des Mammakarzinoms. Zuckschwerdt Verlag (2012)

13. Lenz, R., Reichert, M.: IT Support for Healthcare Processes - Premises, Challenges, Perspectives. Data Knowl. Eng. 61(1), 39–58 (2007)

14. Marcos, M., Roomans, H., ten Teije, A., van Harmelen, F.: Improving medical protocols through formalisation: a case study. In: Proceedings of the 6th World Conference on Integrated Design and Process Technoloy (2002)

15. Müller, H., Schmid, K., Conen, D.: Qualitätsmanagement: Interne Leitlinien und Patientenpfade: Medizinische Klinik, vol. 96(11), pp. 692–697 (2001)

16. Mulyar, N., van der Aalst, W.M.P., Peleg, M.: A Pattern-based Analysis of Clinical Computer-interpretable Guideline Modeling Languages. J. Am. Med. Inform. Assoc. 14(6), 781–787 (2007)

17. National Institute for Health and Clinical Excellence. Early and locally advanced breast cancer. Diagnosis and treatment (2009)

18. Oberender, P.O.: Clinical Pathways: Facetten eines neuen Versorgungsmodells, 1st edn. Kohlhammer, Stuttgart (2005)

19. Peleg, M., Tu, S., Bury, J., Ciccarese, P., Fox, J., Greenes, R.A., Hall, R., Johnson, P.D., Jones, N., Kumar, A., Miksch, S., Quaglini, S., Seyfang, A., Shortliffe, E.H., Stefanelli, M.: Comparing computer-interpretable guideline models: a case-study approach. J. Am. Med. Inform. Assoc. 10(1), 52–68 (2003)

20. Schlieter, H.: Ableitung von Klinischen Pfaden aus Medizinischen Leitlinien Ein Modellbasierter Ansatz. Dissertation, Technische Universität Dresden (2012)

21. Shiffman, R.N., Michel, G., Essaihi, A.: Bridging the Guideline Implementation Gap: A Systematic, Document-Centered Approach to Guideline Implementation. J. Am. Med. Inform. Assoc. 11(5), 418–426 (2004)

22. Vanhaecht, K., De Witte, K., Sermeus, W.: The impact of clinical pathways on the organisation of care processes. PhD dissertation KULeuven (2007)

Semantic Web and Ontology Engineering for the Colorectal Cancer Follow-Up Clinical Practice Guidelines

Hongtao Zhao and Kalpdrum Passi

Dept. of Math. & Computer Science, Laurentian University, Sudbury ON P3E2C6, Canada
{hy_zhao,kpassi}@laurentian.ca

Abstract. Follow-up care for Cancer patients is provided by the oncologist at the cancer center. There are administrative and cost advantages in providing the follow-up care by family physicians or nurses. This paper presents a Semantic Web approach to develop a decision support system for the Colorectal Cancer Follow-up care that can be used to provide the follow-up care by the physicians. The decision support system requires the development of Ontology for the follow-up care suggested by the Clinical Practice Guidelines (CPG). We present the ontology for the Colorectal Cancer based on the follow-up CPG. This formalized and structured CPGs ontology can then be used by the semantic web framework to provide patient specific recommendation. In this paper, we present the details on the design and implementation of this ontology and querying the ontology to generate knowledge and recommendations for the patients.

Keywords: semantic web, ontologies, clinical practice guidelines (CPG), Colorectal Cancer Follow-up.

1   Introduction

Follow-up care is a medical program that (cancer) patients participate in after their primary treatment. Some of the key points are [5]:

a) Follow-up care involves regular medical checkups and tests, which includes a review of patients' medical history and physical exams;

b) The purpose of follow-up care is to check for (cancer) recurrence, the return of cancer in the primary, or metastasis, which is the spread of cancer to another part of the body;

c) The visits during follow-up care may be helpful to identify or address some treatment-related problems patients may have, or to check for problems that continue or may arise after treatment ends;

d) Follow-up care is specified depending on the type of cancer, the type of treatment patients received and patients' overall health condition;

e) Follow-up care can be divided into long-term versus short-term that is determined by the length of the program; or it can be divided into high-intensive and low-intensive program which is judged by the visit or test frequencies;

f) There are usually some guidelines for each follow-up care program, although different doctors may have different emphases in practice.

G. Huang et al. (Eds.): HIS 2013, LNCS 7798, pp. 53–64, 2013.

Follow-up care helps to identify changes in patients' health, early detection of other types of cancer, solve ongoing problems related to the cancer or its treatment. Oncologist discusses the follow-up program with the patient based on the Clinical Practice Guidelines (CPG) which is followed by the patient. For efficient administrative and cost effective treatment, the follow-up can be administered by a physician or a nurse. This can be done by providing the physician or the nurse with a decision support system based on the CPG to help provide the follow-up care at the secondary care level.

Clinical Practice Guidelines (CPGs) [2], [3], [6], [13] for follow-up care is the key both for the physicians as well as patients. It is evidence-based recommendation to use in a more general clinical setting where specialist is not required. In order to give follow-up care, physicians need access to the CPGs and then apply it to individual patient. It is the first document reference for the physicians. CPGs can also be used to standardize medical care, improve quality of care, reduce kinds of risks (mainly to the patient) as well as achieve the balance between cost and medical factors, such as efficiency, specificity, sensitivity and clearness.

In this paper we use semantic web [11] technology to construct well-defined ontology [10] to build a decision support system for the Colorectal Cancer follow-up care based on the CPG. We do not present the complete framework for the decision support system due to limited space. Ontology for the Colorectal Cancer follow-up CPG [2] is created by defining the relevant classes, properties and its relationships.

2   Semantic Web and Ontology

Semantic Web can be referred to the technology of "Web of linked data" that can enable people to create data accessible through the Web, build vocabularies to eliminate the ambiguities, and define rules to handle data.

The Semantic Web component, which is also known as the Semantic Web Stack, consists of XML, XML Schema, RDF, RDF Schema and OWL [8] as its standards and tools. The structure and architecture of this stack is shown in Figure 1. The main components are defined below.

1. RDF is a language used to express data models as resources and the relationships amongst the resources. RDF can be represented in various syntaxes, such as RDF/XML, N3, and Turtle.
2. RDF Schema (RDFS) extends the RDF by adding semantic vocabularies to describe properties and classes of RDF resources.
3. OWL adds more vocabularies to describe properties and classes, such as relations between classes, cardinality (e.g. "some", "exactly one"), equality, richer typing and characteristics of properties and enumerated classes, it improves the semantics in the ontologies.
4. RIF (Rule Interchange Format) is the Rule Layer of the Semantic Web Stack, and
5. SPARQL [12] is a language to query data from Semantic Web sources, such as RDF and OWL.

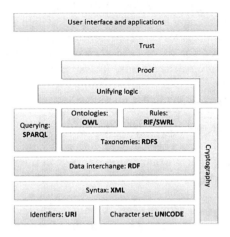

Fig. 1. The Semantic Web Stack

Ontologies are used to represent the knowledge as a set of concepts within a domain and the relationships amongst them. It can also be used to reason about the entities within domains and describe the domains. Ontologies can be used as a form of knowledge representation about things, which makes the ontologies as a very important component in the Semantic Web that provides meaning to the text. Although there are some differences within different formats of ontologies, most of the ontologies consist of the following components:

- Individuals: instances or objects as the entities
- Classes: sets or collections of things, which are categorized by types
- Attributes: properties, characteristics about each class or individual
- Relations: how classes and individuals are related to each other
- Function terms: complex structure, or combination of certain relations
- Restrictions: define what should be true or false to obtain logical inferences
- Rules: if-then statements that describe the logical inferences to obtain assertions
- Axioms: assertions and rules that comprise the overall theory

After all the content of components is created within ontologies, it can be coded into ontology language, such as Web Ontology Language (OWL).

3   Ontology Engineering Colorectal Cancer Follow-Up Care CPG

American Society of Clinical Oncology [1] released an update [4] of Colorectal Cancer follow-up care CPG based on published evidence and results from three independently reported meta-analyses of randomized controlled trials. The abstract of the update (new version) of the follow-up care CPG for Colorectal Cancer is:

1. Annual computed tomography (CT) of the chest and abdomen for 3 years after primary therapy for patients who are at high risk of recurrence and who could be candidates for curative-intent surgery;
2. Pelvic CT scan for rectal cancer surveillance, especially for patients with several poor prognostic factors, including those who have not been treated with radiation;
3. Colonoscopy at 3 years after operative treatment, and, if results are normal, every 5 years thereafter;
4. Flexible proctosigmoidoscopy every 6 months for 5 years for rectal cancer patients who have not been treated with pelvic radiation;
5. History and physical examination every 3 to 6 months for the first 3 years, every 6 months during years 4 and 5, and subsequently at the discretion of the physician;
6. Carcinoembryonic antigen every 3 months postoperatively for at least 3 years after diagnosis, if the patient is a candidate for surgery or systemic therapy.

3.1   Ontology Design

The following classes should be created to represent the CPG as ontology:

1. Patient: the CPG is designed to help deliver follow-up to patients, and every statement of the CPG is about patients;
2. Disease: the CPG is designed to conduct follow-up for Colorectal Cancer, which is a disease. This class can include all the diseases, and in this particular case, has Colorectal Cancer;
3. Follow-up: follow-up will represent as the class of all the follow-up care CPGs, which in this case includes CPG for Colorectal Cancer follow-up;
4. Treatment: all the tests, laboratory exams, and medicines mentioned in the CPGs.

Object (Class) Properties

Next, the object (class) properties have to be defined so that the classes can be connected to one another. The first class is the Disease, so that each disease has some follow-up care. As a convention to ontology naming, *hasFollowup* property is created, where the domain is the class of *Disease* while the range is the class of *Follow-up* that connects the classes as "*Disease* has some *Follow-up*".

Second property would be *involveTreatment* between the classes *Follow-up* and *Treatment* since every follow-up involves at least one test or exam. The two classes can be connected as "*Follow-up* has some *Treatment*".

recommendedFor property connects classes *Follow-up* and *Patient*. There are six items in the Colorectal Cancer follow-up CPG, with different requirements and condition of patients. Each item should be designed for different patients. Thus the classes are connected as "*Follow-up* is recommended for a *Patient*".

OWL supports inverse properties, i.e. the inverse relationship from class A to B as a relationship from class B to A. *isFollowupOf* is the inverse property of *hasFollowup* between the classes *Disease* and *Follow-up*. It represents the relationship "*Follow-up* is follow-up of *Disease*".

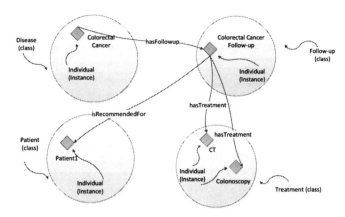

Fig. 2. Classes and Object relationships in the Colorectal Cancer follow-up CPG Ontology

So in all there are four object properties in the CPG ontology to represent the relationships amongst the classes. Meanwhile, these object properties embed semantics into the four classes as shown in Figure 2.

Data Properties

In order to efficiently develop the ontology, data properties are carefully defined according to the CPG. Data properties provide more information such as attributes. Individuals in each class have different data properties, as attributes for each class. After carefully studying the CPG mentioned above, the following data properties are created and associated to individuals in each class:

Patient class:

- *atHigherRiskofRecurrence*: indicates whether a patient is at higher risk of recurrence, with binary value "yes" or "no"
- *isCandidateforCurative-intentSurgery*: specifies if a patient is a candidate for curative-intent surgery, with binary value "yes" or "no"
- *hasSeveralPoorPrognosticFactors*: indicates whether a patient has several poor prognostic factors, with binary value "yes" or "no"
- *hasTreatedwithRadiation*: indicates if a patient has been treated with radiation, with binary value "yes" or "no"
- *hasTreatedwithPelvicRadiation*: indicates if a patient has been treated with pelvic radiation, with binary value "yes" or "no"
- *hasCancerSpecified*: gives more information on the specific disease for the patient, for example, colon is a specific cancer of Colorectal Cancer
- *isCandidateforSurgeryorSystemicTherapy*: indicates if a patient is a candidate for surgery or systemic therapy or not

Disease class:

- *name*: provides formal name of disease

Follow-up class:

- *name*: provides formal name for disease, the same as the data property in the class Disease
- *startFrom*: indicates the start time or date of such follow-up
- *position*: where to proceed the actual test on the patient
- *frequencyWithinFirst3YearsMax*: defines the maximal test frequency within the first 3 years after the follow-up is created
- *frequencyWithinFirst3YearsMin*: defines the minimal test frequency within the first 3 years after the follow-up is created
- *conditionOfFirst3Years*: gives more information of the condition on such follow-up item to the physician to arrange follow-up, within first 3 years
- *frequencyAfterFirst3YearsMax*: defines the maximal test frequency after the first 3 years of the follow-up
- *frequencyAfterFirst3YearsMin*: defines the minimal test frequency after the first 3 years of the follow-up
- *conditionAfterFirst3Years*: gives more information of the condition after the first 3 years on such follow-up item to the physician to arrange follow-up

Treatment class:

- *name*: formal name of test, exam, medicine
- *abbreviation*: abbreviation of test name, such as CT stands for Computed Tomography

The design of the ontology described above has four classes, four object properties and seventeen data properties. Next the designed ontology is coded into OWL.

3.2   Coding the Ontology

In order to code the designed ontology into OWL, protégé [9] is used to accomplish this task. Protégé is a powerful OWL editing tool with GUI using JENA [7] engine of Semantic Web [11]. The very first step is to name the ontology, i.e. the URI, as the URI is the unique address and name for ontology. In the paper, the URI of Colorectal Cancer follow-up CPG Ontology is

http://fccwebsys.cs.laurentian.ca/ontologies/2012/5/colorectalcancerfollowup.owl
This is the URI of the ontology, and with the ontology document placed on the server, everyone can access this ontology through the web.

Protégé provides a friendly interface that allows all operations to the ontology in an efficient and easy manner. Figure shows adding a new class to the Colorectal Cancer Follow-up CPG ontology. In OWL and protégé, everything belongs to Thing, and all the classes are subsets of the Thing. Comments can be easily added into the ontology, which is intended for programmers to know how to use the ontology or the functionality provided by a class. Once the classes and properties are created, they can be accessed by adding # after the URI and names after # to access. For example, once class Disease is created, it can be accessed by visiting
http://fccwebsys.cs.laurentian.ca/ontologies/2012/5/colorectalcancerfollowup.owl#Disease

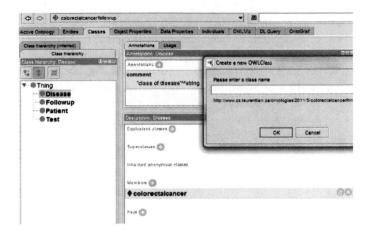

Fig. 3. Add class into ontology in Protégé

Figure shows how to add object (class) properties to the ontology. As can be seen in this figure, object property *hasFollowup* belongs to the domain of *Disease*, while on the other hand it might belong to some other class also. So the keyword "some" is used to indicate that *hasFollowup* has the domain of *Disease*, but is not limited to this class/domain. Class *Followup* is the range of *hasFollowup*, which will include the individuals of actual follow-up items, while *isFollowupOf* is the inverse property of *hasFollowup*.

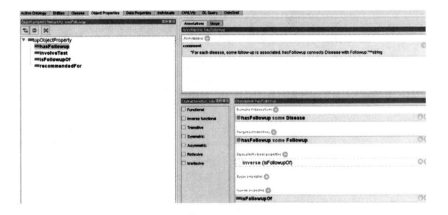

Fig. 4. Add object properties to ontology

Figure shows the screenshot of creating data properties into the ontology. The type or value of data property could be Integer, String, or Boolean as regular data types. As in this ontology, Integer is selected to represent number of months and String to represent text and sentences. Figure 6 shows the individuals and its data properties after the ontology is created. For example the 6th item of the CPG converted to ontology is named as followupitem6. In the object property assertions, it can be seen that this item *involvesTest* carcinoembryonicantigen is one individual in the *Treatment* class. After all the classes, object properties and data properties are created, individuals are required to be added into the ontology to represent actual semantics. The classes and properties are abstract rules, until the individuals are added into the classes, semantics or knowledge cannot be represented. Individuals are added accordingly with divergent data properties most of the time as shown in Figure.

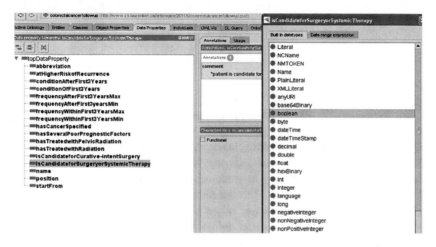

Fig. 5. Creating data property for ontology

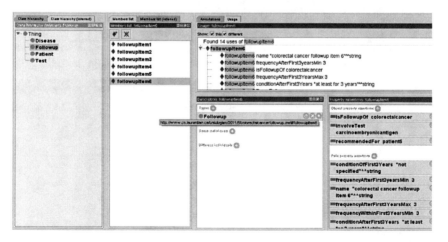

Fig. 6. Individuals in the ontology

3.3   SPARQL Query

SPARQL is an RDF query language and its name is an acronym that stands for SPARQL Protocol and RDF Query. The query consists of triple patterns, conjunctions, disjunctions, and optional patterns. SPARQL allows unambiguous queries from users, unlike SQL, which for example treats both "email" and "e-mail" as the same, and that SPARQL deals with URI, which is defined uniquely. SPARQL defines four different query types (variations) to achieve different goals. They are defined below.

SELECT query: Query to extract raw values from a SPARQL endpoint (through HTTP, or local dataset), the results are returned in a table format. SELECT defines the named variables that are in the result set.

CONSTRUCT query: Query to extract information from SPARQL endpoint and to transform the results into valid RDF. This query builds an RDF on a graph template. In the graph template, named variables are bound by a WHERE clause.

DESCRIBE query: Query to extract an RDF graph from the SPARQL endpoint, while the content is left to the endpoint to decide about the useful information. For each URI found or mentioned in the DESCRIBE clause, the query processor should provide a useful fragment of RDF, such as all the known details.

ASK query: The ASK query returns a Boolean result; true if the pattern is matched, false otherwise.

Each of the query forms above involves a WHERE clause that puts constraints on the query to minimize the results.

```
PREFIX table:
<http://fccwebsys.cs.laurentian.ca/ontologies/2012/5/colorectalcancer
followup.owl#>
SELECT ?name ?diseasename ?position ?starttime ?testname ?hasCancer
FROM
<http://fccwebsys.cs.laurentian.ca/ontologies/2011/5/colorectalcancer
followup.owl>
  WHERE
  {
     ?hasFollowup table:name ?name.
     ?hasFollowup table:isFollowupOf ?disease.
     ?disease table:name ?diseasename.
     ?hasFollowup table:position ?position.
     ?hasFollowup table:startFrom ?starttime.
     ?hasFollowup table:involveTest ?test.
     ?test table:name ?testname.

  OPTIONAL {
          ?patient table:hasCancerSpecified ?hasCancer.
  }
     FILTER(regex(?hasCancer, "rectal") && ?diseasename = "colorectal
cancer").
  }
```

An example of the SPARQL query on the CPG could be written as "what are the follow-up items for the Colorectal Cancer with rectal cancer in particular? Provide the name of each follow-up item, and for each item, provide the name and position of the test, and mention the start time of the test":

As shown above, the first line of the query is to define the prefix. By doing this, it makes the query clean and understandable, and defines "table" to represent the URI of the CPG ontology (owl). Variables are indicated by a "?" or "$" prefix.

OPTIONAL clause is used to take care of undefined properties. For example, in the example above, variable ?hasCancer may not occur in some follow-up items, so this variable or property is optional. If the OPTIONAL clause was not defined, hasCancer information will not be retrieved. OPTIONAL provides extended information, so that if the condition is met in the OPTIONAL clause, the solution will be extended, otherwise the original solution will be returned.

FILTER clause is used to restrict the result to a minimal set. The condition in the FILTER clause must be met, as in an ontology dataset (OWL) a large amount of data may be defined and stored. Using FILTER will get the results that meet certain conditions. For example, in the above query, regex stands for regular expression, that means the value of the variable ?hasCancer must contain the string "rectal", while the value of the variable ?diseasename must be "colorectal cancer". A logical AND "&&" specifies that the two conditions must be true at the same time, so the FILTER clause shows that the follow-up items must belong to the disease of colorectal cancer, and the type of this cancer must be rectal.

The SPARQL query processor will search for sets of triples to match this pattern above, binding the variables in the query to the corresponding parts of each triple.

Results Binding: After sending the query, the corresponding results will be returned in XML format. The SPARQL results document begins with sparql document element in XML, with the namespace of http://www.w3.org/2005/sparql-results#.

There are two sub-elements inside the sparql element, head and results. While for the ASK query, the sub-elements would be head and boolean, the elements must appear in the defined order.

Head: head is the first element in sparql element. Most of the time head must contain a sequence of elements describing the set of the names of query variables in the queried order from the select, or in the corresponding order of the results element, and it is the same order as in the select query, while the order will be undefined if select * is used, which means select everything about the results.

Inside the head element, there are empty child elements with the same name called variable. This variable is provided with an attribute called name, which would have the values of the queried variables.

In the case of a boolean query result, there will be no element inside head element. Sometimes a link element may appear inside the head element with an attribute called link that links to another URI. This would be some additional metadata besides the regular variable element inside the head.

Results: The `results` contain all the query results. While inside the `results` element, individual `result` element is present as a group of result unit. Each `result` element corresponds to one query solution in a result and contains child elements of the query variables that appear in the solution (query).

Each binding inside a solution is coded as an element `binding` as the child element of `result`, with the variables appearing as the value of `name` attributes. The following snippet shows how `results` element looks like when it is binding two variables testname and position:

```xml
<?xml version="1.0"?>
<sparql xmlns="http://www.w3.org/2005/sparql-results#">
<head>
  <variable name="testname"/>
  <variable name="position"/>
</head>
<results>
  <result>
    <binding name="testname">
     <literal xml:lang="en">CT</literal>
    </binding>
    <binding name="position">
     <literal xml:lang="en">Chest</literal>
    </binding>
  </result>
  <result>
    <binding name="testname">
     <literal xml:lang="en">CT</literal>
    </binding>
    <binding name="position">
     <literal xml:lang="en">Pelvic</literal>
    </binding>
  </result>
  ...
</results>
</sparql>
```

As shown above, in the `results` there are two individual `result` units, one is CT at position of Chest, the other is CT at position of Pelvic. For the Boolean results, it would be <boolean>true (or false)</boolean> directly after `head` element.

3.4 Testing the Ontology

After the Colorectal Cancer follow-up Ontology is created, SPARQL has to be implemented to test the outcome of the query result. In the development of the CPG ontology engineering, twinkle [14] is used as a SPARQL tool to query the OWL ontology. A SQL-like query against the ontology can generate all the information required to represent the CPG knowledge. As tested, it provides the expected results as desired, and can fully support the development of the proposed framework.

4   Conclusions

In this paper we demonstrated the use of Semantic Web technology to develop the Colorectal Cancer follow-up CPG ontology and to query the ontology. A decision support system has been created using this ontology as a knowledge source to generate follow-up recommendations for the patients although it has not be presented in this paper due to lack of space. The decision support system can help physicians to decide the details of the follow-up for individual patients, to make decision on test selection, frequency selection and so on. The decision support system is designed to be as easy to use as a regular web application, but has the underlying semantics to generate knowledge.

References

1. ASCO, http://www.asco.org/ (accessed June 1, 2012)
2. Burgers, J.S., Grol, R., Klazinga, N.S., Mäkelä, M., Zaat, J.: For the AGREE Collaboration. Towards evidence-based clinical practice: an international survey of 18 clinical guideline programs. Int. J. Qual. Health Care. 15, 31–45 (2003)
3. Council of Europe. Developing a methodology for drawing up guidelines on best medical practice. Recommendation Rec (2001) 13 and explanatory memorandum. Council of Europe Publishing, Strasbourg (2002)
4. Desch, C.E., Benson, A.B., Somerfield, M.R., Flynn, P.J., Krause, C., Loprinzi, C.L.: Colorectal cancer surveillance: 2005 update of an American Society of Clinical Ontology practice guideline. J. Clin. Oncol. 23, 8512–8519 (2005)
5. Follow-up, http://www.cancer.gov/cancertopics/factsheet/Therapy/followup (accessed January 10, 2012)
6. Institute of Medicine (edt.) Clinical practice guidelines we can trust. Washington DC (2011)
7. JENA: Semantic Web Framework, http://jena.sourceforge.net/documentation.html (accessed July 18, 2012)
8. OWL Web Ontology Language Overview, http://www.w3.org/TR/owl-features/ (accessed July 10, 2012)
9. Protégé, http://protege.stanford.edu/ (accessed July 1, 2012)
10. RDF, http://www.w3.org/RDF/ (accessed June 1, 2012)
11. Semantic Web, http://w3.org/standards/semanticweb/ (accessed January 21, 2012)
12. SPARQL, http://www.w3.org/TR/rdf-sparql-query/ (accessed August 1, 2012)
13. The AGREE Collaboration. Development and validation of an international appraisal instrument for assessing the quality of clinical practice guidelines: the AGREE project. Qual. Saf. Health Care. 12, 18–23 (2003)
14. Twinkle, http://www.ldodds.com/projects/twinkle/ (accessed August 1, 2012)

Neural Network Solution
for Intelligent Service Level Agreement in E-Health

Nada Al Salami and Sarmad Al Aloussi

Abstract. In the next twenty years, service-oriented computing will play an important role in sharing the industry and the way business is conducted and services are delivered and managed. This paradigm is expected to have major impact on service economy; the service sector includes health services (e-health), financial services, government services, etc.With increased dependencies on Information and Communications Technology (ICT) in their realization, major advances are required in user (Quality of Services) QoS based allocation of resources to competing applications in a shared environment provisioning though secure virtual machines.

In this paper, we pointed in addressing the problem of enabling Service Level Agreement (SLA) oriented resources allocation in data centers to satisfy competing applications demand for computing services. e-Health offers a QoS Health Report designed to compare performance variables to QoS parameters and indicate when a threshold has been crossed. e-Health graphs relevant performance metrics on the same axes as thresholds indicative of SLAs or equivalent requirements. We suggest a methodology which helps in SLA evaluation and comparison. The methodology was found on the adoption of policies both for service behavior and SLA description and on the definition of a metric function for evaluation and comparison of policies. In addition, this paper contributes a new philosophy to evaluate the agreements between user and service provider by monitoring the measurable and immeasurable qualities to extract the decision by using artificial neural networks (ANN).

Keywords: Service Oriented Architecture, Service Level Agreements, e-Health, a QoS, and Neural Network.

1   Introduction

The term e-health, referring to all digital health-related information, in the latter part of the nineteenth and early part of the twentieth century, medical applications were quick to derive benefit from the progress being made in the field of analogue telephony. The technology enabled not only individuals to call the doctor but also hospitals to transmit electrocardiograms over telephone lines. Digital telemedicine has experienced tremendous growth over the past 25 years and is now a major component of e-health. It enables, among other things, the exchange of healthcare and administrative data and the transfer of medical images and laboratory results. The prefix "e-", standing for "electronic", is similarly used in numerous other applications such as "elearning", "e-governance" and "e-transport", to convey the notion of digital

G. Huang et al. (Eds.): HIS 2013, LNCS 7798, pp. 65–77, 2013.

data. Without digitization there would be no automatic processing and no instantaneous exchange via the network. The term "health" is used broadly and does not refer solely to medicine, disease, healthcare or hospitals. The scope of e-health is health in general, with its two major facets, namely public health –which is the responsibility of States and is geared towards preventing and responding to disease in populations – and healthcare, which is geared towards E-health products, systems and services are location independent, in that they can be used locally (doctors' surgeries, hospitals) or remotely, as is inherent in the term "tele" (teledermatology, telesurgery, telediagnosis and so on). Currently, health information is accessed from and transferred over many different types of computers, telecommunications networks and information systems .Often these have been implemented in isolation of one another making it difficult and costly to share information between providers and systems in a secure way. In a person-centered health system the ability to connect services, applications and systems is essential for allowing patients to be cared for by the right health provider, at the right time and place, providing access to patient records electronically with the confidence that information is kept secure at all stages. The Connected Health Program is a key step in achieving this aim. Its purpose is to establish the secure environment needed for the safe sharing of health information between all the participating health providers.

Service-level agreements are, by their nature, "output" based — the result of the service as received by the User is the subject of the "agreement." The (expert) service provider can demonstrate their value by organizing themselves with ingenuity, capability, and knowledge to deliver the service required, perhaps in an innovative way. Organizations can also specify the way the service is to be delivered, through a specification (a service-level specification) and using subordinate "objectives" other than those related to the level of service. This type of agreement is known as an "input" SLA.

The cooperation between services and service oriented architectures for e-health need to interact between them, generally service level agreements expressed in ambiguous ways and this implies that they need to evaluate both in a mutual agreement to qualify a service and in monitoring process.

The development of SOA, organization is able to compose complex applications from distributed services supported by third party providers. Service providers and User negotiation based service level agreement (SLA) to determine different activities (security, cost, penalty,.....etc) on the achieved performance level. The service providers need to manage their resources to maximize the profits.

To maximize the SLA revenues in shared data environments, it can be formulated as the dual problem of minimizing the response time and maximizing throughput. That proposal considers the problem of hosting multiple web sites.

2   Policies to Express Service Level Agreements

We need a formal way to express Health Service Levels and automate the Agreement process (sick and system). To do this we will adopt a policy way for the formalization and an evaluation methodology for the automatic agreement. Policies can be expressed and formalized in mathematical formal or intelligent formal. We could classify the following types of policies:

1. Formal policies are usually expressed in mathematical form.
2. Semi formal policies are partially expressed in artificial form.

The formal policies are typically expressed by us according to service provider model need to express in an unambiguous way technical procedures, while the needing to express practical and behavioral aspects of the service providing process for secure form policies. Both technical and organizational aspects are very critical, the policy formalization and the evaluation process are performed by an artificial model. It intends to make the decision more sensitive to the policy formalization in terms of what to express in a security policy and in which formalism to express it [1],[2].

3   The Evaluation Methodology

Having formalized and expressed SLAs by a policy form, in this thesis we need an evaluation methodology to compare them and decide to request a service from the server or changing the server to provide the request service. The proposed methodology is based on a Reference Evaluation Model (REM) to evaluate and compare different security policies and behavior policies, quantifying their security level by either value 0 (represent not existing sub provision for each immeasurable provisions) or value 1 (represent existing provisions) . The model will define how to express in a rigorous way the security policy (formalization), how to evaluate a formalized policy, and what is its service level. In particular the REM is made of three different components:

1. The policy formalization,
2. The evaluation technique,
3. The reference levels.

3.1   The Policy Formalization

The formalization policy is a way to express all the qualities parameters for SLA in a rigorous technique to define either the qualities exist or not, or for measurable qualities with defined values.

These qualities can be applied directly to evaluate it either mathematically or artificially to extract the suitable decision.

3.2   The Evaluation Technique

We propose REM includes the definition of a technique to compare and evaluate the policies; we have called this component the REM Evaluation technique. Different evaluation techniques represent and characterize the measurable and immeasurable level associated to a policy in different ways, for example with a numerical value, a fuzzy number [3] or a verbal judgment representing its security level.

3.3   The Reference Levels

The last component of the REM is the set of SLAs levels that could be used as a reference scale for the numerical evaluation of SLA. When references are not available, the REM could be used for direct comparison among two or more policies.

4   SLA Structure

The service providers and the users often negotiate by the qualities to base the Service Level Agreements (SLAs) by behavior aspects based on the achieved performance levels. The service provider needs to manage its resource to maximize its profits. The optimization approaches are commonly used to provide the service load balancing and to obtain the optimal classifications for quality of service levels. By the above are also used as guidelines and for realizing high level trends. One main issue of these systems is the high variability of the workload according to values of measurable and immeasurable qualities [4].

By such model, the service can dynamically be allocated among the service providers depending on the service availability. Fig.1 shows the architecture of SLA model implementing an autonomic infrastructure. Service providers are allocated and de-allocated on demand on servers. The server level agreement model can monitor the qualities and by predictor phase, it can allocate the server to provide the service.

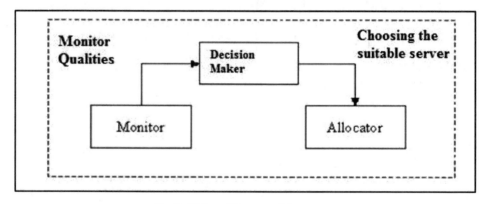

Fig. 1. SLA architecture of data center

The main components of the SLA model [2] include a monitor, a decision maker and a server allocator. The system monitors the qualities and performance matrices of each form, identifies requested classes and estimates requested service time. The decision maker can evaluate the system performance from the trace values. The allocator chooses the best system configuration.

5   Service Level Agreement Categories

SLA was defined as a contract between the users and service providers, and then the contract has many rules. Their rules can be established by different ways either manually or automatically. It can be either static, which means the system will be fixed without any modifications or upgrading all the service providing or the contract rules be dynamic and changing all the time, therefore SLAs typically fall into three categories:

Basic: a single SLA with well established matrices those are measured and/or verified. The collection of these matrices is typically done manually.

Medium: the introduction of multi-level quality based on the cost of the service. The objective is to balance the levels of quality and cost. Automatic data collection enables comprehensive reports.

Advanced: dynamic allocation of resources to meet demand as business needs evolve.

From all the above, the SLA can be classified according to the inputs values collection way and the SLA, the inputs can define as qualities can classify as below:

5.1   Measurable Qualities

There are many measurable qualities; it can measure for each user, the definitions of the measurable qualities are shown below:

Accuracy is concerned with the error rate of the service. It is possible to specify the average number of errors over a given time period.

Availability is concerned with the mean time to failure for services, and the SLAs typically describe the consequences associated with these failures. Availability is typically measured by the probability that the system will be operational when needed. It is possible to specify the system's response when a failure occurs – the time it takes to recognize a malfunction.

Capacity is the number of concurrent requests that can be handled by the service in a given time period. It is possible to specify the maximum number of concurrent requests that can be handled by a service in a set block of time.

Cost is concerned with the cost of each service request. It is possible to specify the cost per request the cost based on the size of the data – cost differences related to peak usage times.

Latency is concerned with the maximum amount of time between the arrival of a request and the completion of that request.

Provisioning-related time (e.g., the time it takes for a new client's account to become operational).

Reliable messaging is concerned with the guarantee of message delivery. It is possible to specify how message delivery is guaranteed (e.g., exactly once, at most once) whether the service supports delivering messages in the proper order.

Scalability is concerned with the ability of the service to increase the number of successful operations completed over a given time period. It is possible to specify the maximum number of such operations.

5.2   Immeasurable Qualities

There are three main immeasurable qualities that can be defined by main provision and sub provision for each quality; in the following we define the immeasurable qualities:

Interoperability is concerned with the ability of a collection of communicating entities to share specific information and operate on it according to an agreed upon operational semantics. Significant challenges still need to be overcome to achieve semantic interoperability at runtime.

Modifiability is concerned with how often a service is likely to change. It is possible to specify how often the service's

Interface changes.

Implementation changes.

Security is concerned with the system's ability to resist unauthorized usage, while providing legitimate users with access to the service. Security is also characterized as a system providing non-repudiation, confidentiality, integrity, assurance, and auditing. It is possible to specify the methods for

authenticating services or users
authorizing services or users
encrypting the data

5.3   Policy Formalization

It will depend on measurable and immeasurable qualities which are mentioned is previous section. It will form two unsymmetrical matrices, first one describes the immeasurable qualities for n users, as well as the output matrix describe the decided output. The second matrix represent the measurable matrix, each row represent the provision and the state of each column represent if the sub provision exist or not, the output matrix represent the trace value of measurable matrix. Trace is the value of each matrix by maximum value of matrices.

	User1	User2	User3	-------------------------------------	Usern--1	Usern
Accuracy	-	-	-		-	-
Availability	-	-	-		-	-
Capacity	-	-	-		-	-
Cost	-	-	-		-	-
Related Time	-	-	-		-	-
Scalability	-	-	-		-	-

Measurable Matrix

Output usern = [Trace, Request, Delay Request, Wait, Change Provider, New User]

	Prov.1	Prov.2	Prov.3	---------------------------	Prov.m--1	Prov.m
Interoperability	-	-	-		-	-
Modifiability	-	-	-		-	-
Security	-	-	-		-	-

Immeasurable Matrix

Output= [Trace, Activated output]

5.4   Mathematical Evaluation Technique

To adopt WS-policy framework and to express policies for security-SLA the framework is structured as a hierarchical tree to express all sub provisions. We have started the formalization by considering the set of items proposed by [5]; the first level of the tree structure includes:

1. Security documentation,
2. Security Auditing,
3. Contingency Planning,
4. User Security Training,
5. Network Infrastructure Management,
6. Physical Security,
7. User Discretionary Access Control (DAC) Management,
8. Password Management,
9. Digital Certificate Management,
10. Electronic Audit Trail Management,
11. Security Perimeter or Boundary Services,
12. Intrusion Detection and Monitoring,
13. Web Server Security,
14. Database Server Security,
15. Encryption Services,

6   Configuration Management

All the above items represent general categories (both technical and organizational) and they are actually expressed in natural language. Second level provisions try to describe all the details about all sub provisions and they express objects that are still complex but bring a more bounded security information; for example the Digital Certificate Management provision includes: Key Pair Generation, Key length, Private Key Protection, Activation Data, Key Controls and verification, Network Security Control, Cryptographic module engineering controls [6].

The provisions defined in the first two steps by either existing or not, the structures of tree nodes identify complex security provisions, leaves identify simple security provisions. Furthermore the other immeasurable qualities Interoperability and Modifiability can be represented in the same way for security provisions.

7   ANN Evaluation Technique and Concepts

A web service can be described broadly as a service available via the web that conducts transactions. E-businesses set up Web Services for clients and other Web Services to access. They have a uniform service provider locator at which they can be accessed and have a set of activated outputs. These web services interact to each other they would need to create and manage service level agreements amongst each other. SLA management involves the procedure of signing SLAs thus creating binding contracts, monitoring their compliance and taking control actions to enable compliance.

To define and develop new intelligent approach that provides optimal services to the end-user in client-server environment. The proposed approach will be used to test the service level agreement by selecting policies formalization for each end user. Artificial neural networks (ANNs) are used as interface between the client side and providers to decide if the service query must send by the current server or there is a need to change the server. In this case a heuristic search algorithm based on genetic algorithm will be run to find good such server. To design the ANN service level agreement (SLA), matrices technique are used by define different policies formalization in different category then it will find the trace for each matrix by maximum value of the matrices, so that it can select the optimum policy formula.This is particularly useful in applications where the complexity of the data or task makes the design of such a function by conventional technique is impractical. Application areas include system identification and control (process control), and decision making. In service level agreement we select different patterns of policy formalization, then train ANNs, so that it can select the optimum policy formula, this formula represent the best formula to test the service behavior in SOA. Fig.2 shows the general concepts of Neural Network Service Level Agreement in SOA.

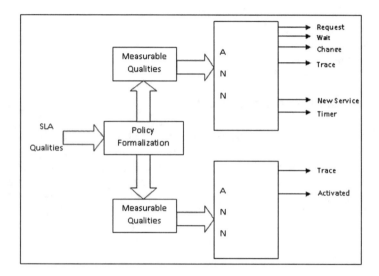

Fig. 2. ANN Service Level Agreements

7.1   Proposed ANNSLA Model

The main idea of this model is to be able to decide if the services can supply by service provider or not depending on the value of the activated outputs. Because of this model was proposed depending on artificial neural network then the ANN needs to train by input and output data sets to be able to extract the outputs for other input data sets. The training input and output data sets will be represented according to the mathematical models. The input sets of ANN represent the measurable and immeasurable matrices which are used in mathematical model while the output sets represents the extracted outputs for the traces and the activated outputs from the mathematical calculations model. There are two phases should execute to extract the correct decision. In the following we will discuss the phases of ANNSLA model:

ANN Training phase: in this phase, the extracted data sets (formalized inputs and calculated outputs) from the SLA mathematical model will apply to ANN to train it as Fig.4. Operation Phase: when the ANN trained correctly according to inputs/outputs data sets, it will be ready to use it as decision maker for different input sets to extract the trace value and activated output or outputs for either immeasurable qualities or measurable qualities. To evaluate the output values, it can compare the extracted outputs with calculated outputs from mathematical model, Fig.4 shows the general trained ANNSLA model. By this model can extract the outputs for different inputs by applying it to trained ANN. As we mentioned in the training phase, the trained ANN be as a decision maker to extract the outputs. When applying the immeasurable or measurable matrices to trained ANN it can extract the trace and activated output or outputs.

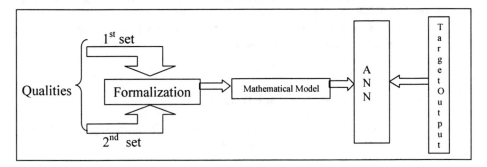

Fig. 3. ANNSLA training phase

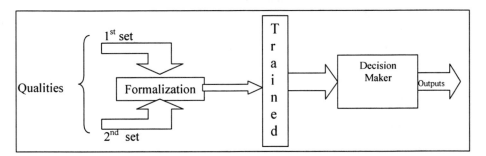

Fig. 4. ANNSLA operation phase

7.2   ANNSLA Process and Simulation

The ANNSLA will build using ANN architecture, it's very important to select the architecture by defining the number of layers (hidden layers), the number of input nodes in input layer, the output nodes in output layer, the values of inputs, the values of target outputs, and the error value. These architectures can be done by using MATHLAB package version 7.7.0. ANNSLA MATLAB Architecture

As we mentioned in the above sections the ANNSLA model architecture has two part one for measurable qualities and the second one for immeasurable qualities, this means there are two ANN.

The measurable ANN part consists the following:

Input layer with consist 6 input neurons for each user. The inputs represent the measurable qualities which are the accuracy, availability, capacity, cost, related time, and scalability. The values of these inputs represent numeric value all the time $\leq 1[6]$.

One or two hidden layers to test the effect of increasing the hidden layers upon the performance of the architecture.The output layer consist either one neuron if we want to calculate trace of measurable matrix or five neurons represent the outputs (Request, Wait, Change, New Service, and Timer) for measurable part in ANNSLA model.

The immeasurable ANN part consists the following:

Input layer with consist 40 input neurons for each user. The inputs represent the immeasurable qualities represent interoperability, modifiability, and security. All the values of these inputs will be represented either 1 (exist) or 0 (not exist).

One or two hidden layers to test the effect of increasing the hidden layers upon the performance of the architecture.

The output layer consist either one neuron if we want to calculate trace of immeasurable matrix or one neuron represents the activated output for immeasurable part in ANNSLA model.

7.3   ANNSLA MATLAB Simulation Data Sets

To train the measurable and immeasurable ANN, we need to prepare the following types of data to be ready when we will train the ANN:

Trace calculation, for both measurable and immeasurable matrices will be calculated according by mathematical forms in MATHLAB package. The mathematical forms represent the trace and the distance equation which are mentioned in mathematical evaluation technique section.

Measurable simulation data sets:

The inputs measurable matrices are represented in 12 input data sets, each matrix represents 5 users.

The outputs matrices are represented in 12 output data sets, these sets represent the output for the measurable ANN part in SLA model. These outputs values represent the target output for the immeasurable ANN.

Immeasurable simulation data sets:

The inputs immeasurable matrices are represented in 17 input data sets, each matrix represents 5 users.

The outputs matrices are represented in 17 output data sets; these sets represent the trace values and activated outputs for the immeasurable ANN part in SLA model. These outputs values represent the target output for the immeasurable ANN.

The above measurable and immeasurable data sets will apply to ANNs to train it. When the ANN trained according to acceptable error values then this trained ANN can be used to extract the actual outputs and be as a decision maker when applying new data sets of measurable and immeasurable qualities for SLA.

ANNSLA MATLAB Training and Running process

The proposed ANN for both measurable and immeasurable parts should train before run it. The training processes will discuss in the next sections: The MATHLAB shows a neural network performance window which mentioned all the setting factors and performance factors. The following figure shows the neural network training performance for measurable part in ANNSLA there are many factors are defined. The main factor is the Epcho (maximum number of iterations); the maximum value of iterations is 1000000. The second factor is the performance (the acceptable error between the target outputs and actual extracted outputs), the acceptable error setting is 0.21. Below the neural network training performance for measurable part in ANNSLA; from this performance window, we can conclude that the actual extracted output reached to the target output by performance value about 0.0607, it is less than the setting value of the gradient. The ANN was trained in about 4211 iterations. The training time is about 46 seconds. The architecture of ANN is 6 inputs in input layer and 5 outputs in output layer with two hidden layers. The error rule is mean square error to calculate the gradient value between the actual and target outputs. The training process was done by MATHLB software packages by writing a simulation program represents all the inputs/outputs values for the measurable qualities and activated outputs. The training process is dependent on actual values then the results of this process will be extracted correctly, if all the trained values are correct otherwise all the results and the process will run in not suitable way this issue considers as a restricted condition to build and define the ANNSLA model for measurable, immeasurable, and trace ANN parts. From all the above the ANNSLA will build depending on the mathematical model execution and results.

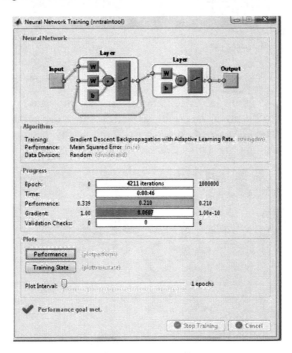

In the same above way, the immeasurable ANN part will train. The setting of the training factors is similar values in the above (maximum number of iteration is 1000000, and the performance value is 0.174.The ANN was trained in about 20321 iterations, the actual extracted output reached to the target output by performance value about 0.00175, and the training time is about 4 minutes and 20 seconds. The figure below shows the performance window for immeasurable ANN part for ANNSLA, it shows all the training parameters. It is very clear the values of the training process parameters.

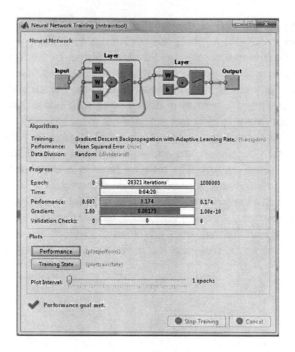

Below the Figure shows the trace training window, the value of Epcho (maximum number of iterations) is about 1697, the second factor is the performance is 0.18, the actual extracted output reached to the target output by performance value about 0.0191,and the training time is about 22 seconds.

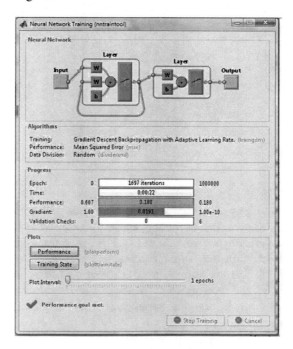

From all the above we conclude that the ANNSLA model can trained in phase part within acceptable error, then the model can extract the actual output/outputs for any measurable and immeasurable data set within minimum performance time.

The trained measurable and immeasurable ANN will train to extract the actual outputs by applying different measurable and immeasurable data sets. For measurable ANN part, we will test 3 data sets and for immeasurable ANN part, we will test 20 data sets.

8 Conclusion

In this paper we have introduced a theoretical methodology to evaluate Service Level Agreement in SOA. The methodology is based on two fundamental features; the first one is the SLA formalization through the use of standard policy while the second one is the formalization of "qualifiable service levels" against which we could measure the SLA.

In particular, we have adopted a Reference Evaluation Model, developed for different methodology, to evaluate and compare different policies and quantifying their levels. The application of the methodology in different samples of measurable and immeasurable qualities and we adopted it in the integration of mathematical model and artificial model to guarantee the same perceived service level to the end-user.

References

1. Kertesz, A., Kecskemeti, G., Brandic, I.: An SLA-based resource virtualization approach for on-demand service provision. In: VTDC 2009, June 15, pp. 27–33. ACM, Barcelona (2009)
2. Abrams, C., Roy, W.: Service-Oriented Architecture Overview and Guide to SOA Research, G00154463. Gartner Research, Stamford (2008)
3. Wetzstein, B., Karastoyanova, D., Leymann, F.: Towards Management of SLA-Aware Business Processes Based on Key Performance Indicators. In: BPMDS 2008, Institute of Architecture of Application Systems, University of Stuttgart, Germany. Springer (June 2008)
4. Bishop, M.: Developing Web Services. In: Proceedings 17th International Conference on Data Engineering, pp. 477–481. IEEE Computer Society, Los Alamitos (2003)
5. Briand, L., Labiche, Y., Shousha, M.: Stress testing real-time systems with genetic algorithms. In: GECCO 2005 Proceedings of the 2005 Conference on Genetic and Evolutionary Computation, pp. 1021–1028. ACM, Washington DC (2005)
6. IBM, WS-policy specification Web Services Policy Framework, IBM, BEA Systems, Microsoft, SAP AG, Sonic Software, VeriSign (2006)
7. Caola, V., Mazzeo, A., Mazzocca, N., Rak, M.: A SLA evaluation methodology in Service Oriented Architectures. Advances in Information Security 23, 119–130 (2006)

Taking Compliance Patterns
and Quality Management System (QMS) Framework
Approach to Ensure Medical Billing Compliance

Syeda Uzma Gardazi and Arshad Ali Shahid

Department of Computer Science
National University of Computer & Emerging Sciences
Islamabad, Pakistan
uzma.gardazi@gmail.com, arshad.ali@nu.edu.pk

Abstract. The United States Office of Inspector General (OIG) has issued a number of compliance guidelines including third-party medical billing guidelines for healthcare companies in the United States to reduce errors and fraud in the field of medical billing. OIG strongly suggests that medical companies should ensure compliance of these guidelines by incorporating OIG requirements within their processes. Compliance of OIG guidelines will ultimately support medical companies to reduce the consequences and penalties of legal actions in case of non-compliance/government audit. This research has suggested that medical companies can adopt an ISO 9001 framework to ensure compliance with OIG billing guidelines. The authors compared ISO 9001 standard with OIG third-party medical billing guidelines by mapping their components. The comparison reveals that the requirements mentioned in the ISO 9001 guideline meet or exceed the OIG third-party medical billing guideline by more than 70 percent. The authors also emphasize the importance of creating additional patterns, named as compliance patterns for the healthcare industry. A few billing compliance patterns are also devised to ensure billing compliance. The United States based third-party Medical Billing Company (MBC) with a backup office in Pakistan was used as a case study to evaluate the effectiveness of this proposed framework. MBC was able to increase customer satisfaction from 72 percent to 84 percent by implementing the proposed framework and compliance pattern approach. Two internal audits were also carried out to evaluate the effectiveness of the proposed framework. It was also clear from the evidence that with a prior implementation and tailoring of the framework approach customer satisfaction tremendously increased.

Keywords: Medical Billing Compliance Patterns, Common Audit Framework, ISO 9001, Quality Management System (QMS), and OIG.

1   Introduction

It is essential for billing companies in the United States to ensure compliance with specific standards and guidelines that restrict illegitimate medical billing activities.

G. Huang et al. (Eds.): HIS 2013, LNCS 7798, pp. 78–92, 2013.

Controlling bodies, e.g., Center of Medicare and Medicaid (CMS), carried out a number of audits to identify, correct, and prevent illegitimate billing activities.

1.1 OIG Guidelines for Third-Party Medical Companies

The Office of Inspector General (OIG) is an independent organization established by the Congress to identify, correct, and prevent billing errors and frauds. The OIG has issued a series of voluntary medical billing guidelines for the United States healthcare industry, including third-party medical billing guidelines to monitor and limit unauthorized medical billing activities. Medical companies can tailor these guidelines to effectively improve internal controls to ensure compliance with applicable standards.

This paper focuses on third-party medical billing guidelines issued by (Federal Register Vol. 63, No. 243)[1]. The term third-party medical billing refers to obtaining the medical billing services of another company or in other words outsourcing billing services. Third-party medical billing companies are also covered under the umbrella of the OIG. These third-party medical companies and their unauthorized billing practices can potentially have a negative impact on the United States healthcare industry. In order to avoid this and to improve third-party medical billing standards, OIG and Department of Health and Human Services (HHS) published medical billing guidelines for third-party companies. The basic purpose of this program is to increase compliance in health care industry throughout the United States, by preventing the submission of incorrect or falsified payment claims to federal health care programs.

Compliance with OIG guidelines is voluntarily, but in case of non-compliance the companies have to face legal action as defined by the United States government. It is highly recommended that healthcare companies should ensure compliance with OIG guidelines to prevent fraud otherwise they will face the consequences of non-compliance.

1.2 ISO 9001:2008 Quality Standard

The International Organization for Standardization (ISO) is an organization responsible for developing and publishing international standards. ISO 9001 is a voluntary standard which specifies generic requirements for quality management systems (QMS) to enhance customer satisfaction and an organization's compliance with applicable standards and regulations.

ISO 9001 standard (reference number ISO 9001:2008 (E)) can be adopted by any type of organization to demonstrate ability to provide quality products [2]. To achieve compliance with ISO 9001 standard, organizations are responsible for incorporating such requirements within their processes. If any requirement(s) of ISO 9001 cannot be implemented for a valid reason, organizations are required to provide evidence for these exclusions and prove that they will not affect the organization's ability to offer quality products to its customers or to comply with applicable standards and regulations.

1.3 Motivation

Recently the trend for outsourcing information, including medical billing, has increased tremendously. Sometimes third-party backup offices work in countries such as Pakistan to utilize the easily available human resource. The United States healthcare professionals also outsource medical billing to these companies after ensuring compliance with applicable regulatory and standard requirements and fulfill the government regulations.

This paper suggests a new approach to ensuring compliance with regulatory requirements using the ISO 9001 framework which can be used by any third-party medical company.

2   Observations and Analysis

The billing operation of a Medical Billing Company (MBC) backup office in Pakistan was taken as a case study. MBC provides medical billing along with IT services to the United States healthcare professionals as third-party. It is essential for MBC to apply a framework to ensure compliance with OIG 3rd party medical billing guidelines. We have done survey and analyses of the MBC to shed light on the importance of Billing Compliance.

2.1   Customer Complaint Survey

A customer complaint survey was conducted in four quarters of the year to see customer satisfaction trend. Seventy five percent customer complaints were related to MBC billing services as shown in Table 1 and figure 1.

Table 1. Customer Complaint analysis

Type	Total	Quarter 1	Quarter 2	Quarter 3	Quarter 4
Billing	75%	417	263	341	170
IT	4%	13	22	24	11

Based on this analysis it can be interpreted that customers face problems/issues in MBC medical billing services.

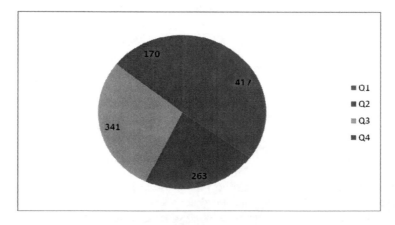

Fig. 1. Quarterly billing complaints analysis

2.2   Billing Compliance Survey of Providers

It is important to identify potential issues in MBC's billing process. For the billing compliance survey that we conducted that contains the following sections [3]:

a) Never visited the doctor: In this case the Patient did not visit the doctor "Provider". The Patient was being wrongly charged.

b) Overcharged: In this case the Patient has to pay more than permitted/ allowed amount.

c) Overpayments: In this case payment was being paid more than allowed amount. It is the responsibility of the Provider to repay the overpaid amount to either Patient or Insurance.

d) The patient was mistakenly sent to the collection agency.

e) Procedure codes were inconsistent with Patient age.

f) The incorrect tax ID number was used to claim.

g) Incorrect fee schedule was selected for payment.

h) Incorrect payer/insurance address was mentioned in the claim.

i) Incorrect Current Procedural Terminology (CPT) or CPT units were mentioned in the claim.

j) The incorrect policy number was mentioned in the claim.

k) Incorrect place of service (POS) was mentioned in the claim.

Analysis: Figure 2 shows statistics of issues being reported by the Providers against billing compliance category.

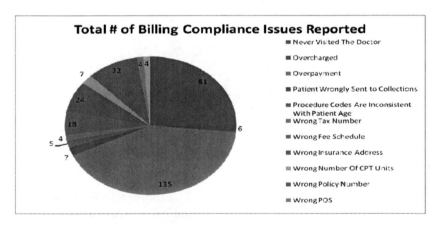

Fig. 2. Total number of billing compliance issues reported

Based on the survey, following top five issues reported are listed as follows:

a) 41 percent issues were about use of incorrect CPT or CPT units.

b) 25 percent issues were categorized as never visited the doctor category.

c) 10 percent issues were categorized as Patient wrongly sent to the collection agency.

d) 7 percent issues were categorized as the wrong insurance address category.
e) 6 percent issues were categorized as the wrong fee schedule category.

Based on this analysis, it can be concluded that the MBC billing process needs to be improved.

2.3   Questionnaire Survey by MBC Employees

Billing survey was arranged for billing teams working in MBC as employees using Microsoft collaboration server. The following issues were addressed in the survey:

- Inappropriate incident-to billing,
- Following verbal instruction(s),
- Using modifiers without Provider's instruction, and
- Receiving unmatched electronic remittance advice (ERA) for Claims not submitted by MBC.

65 percent surveys were completed by billing teams and 35 percent surveys were not fully completed.

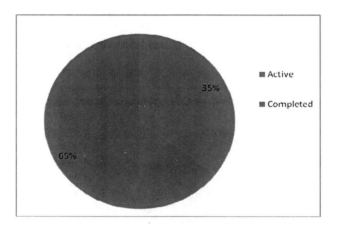

Fig. 3. Questionnaire survey analysis

The survey's first question was related to an incident-to billing issue. The term incident-to is defined by the Medicare to define a situation in which Nurse Practitioner (NP) or Physician Assistant (PA) can perform medical services on an established patient in the presence of physician on site. This claim will be billed under Supervising Physician national provider number (NPI) and will possibly result in 100% payment of the allowed amount defined in Part B Physician Fee Schedule. Incorrect billing is not authorized by the OIG. Currently, 4% inappropriate incident-to billing issues were highlighted by MBC's employees that were created by Providers.

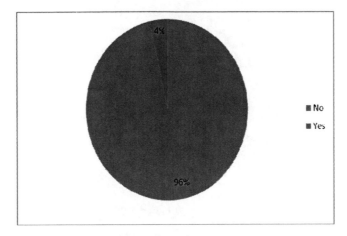

Fig. 4. Inappropriate incident-to billing

The survey's second question was related to verbal instructions. MBC employees receive a number of instructions related to billing. According to company policy, MBC employees should only follow documented and approved instructions by the MBC Compliance Department. Currently 3 percent billing teams were following Provider's verbal instructions.

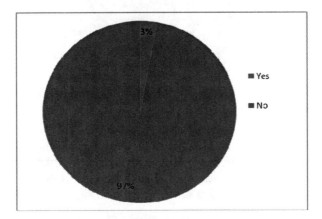

Fig. 5. Following verbal instructions

The survey's third question is related to modifier usage without instructions. MBC employees cannot use modifiers in claims without explicit approval from Providers. Currently, 2 percent billing teams were using the modifier without Provider's written instructions.

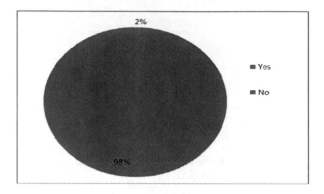

Fig. 6. Using modifiers without Provider's instructions

The survey's fourth question was related to unmatched ERA.

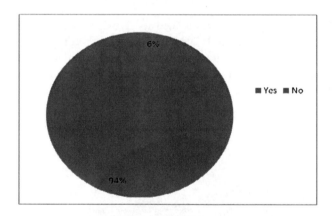

Fig. 7. Received unmatched ERA for Claims not submitted by MBC

The Provider agreement with MBC limits the Providers to submit the claims on their own as they have outsourced billing to MBC. Providers provide the billing information to MBC via either FTP or MBC's Electronic Health Records. Currently 6 percent claims were directly submitted by Providers which are in direct violation of MBC agreement.

2.4   MBC's Existing Billing Audit Process

MBC has laid down and implemented a monthly billing compliance audit plan which highlights the practices ("Providers") and billing teams ("Employees") requirements that need to be audited. The audit has been divided in three phases called "pre-audit", "audit" and "post-audit". The details of each phase are listed below:

a) Pre-audit
   A team select and audit practice based on customer complaints, management instructions and practice compliance health.

The Monthly Practice Compliance Health Report is formulated by MBC's internal audit teams to check current billing compliance status of the billing process. This report is obtained by weighting each of the OIG factors and points are allocated to each practice using the defined percentages below:

Table 2. Practice Selection Criteria

Category	Claim Selection
SSC Complaints	30%
OIG Factors	70%
Practice Compliance Health	**100%**

b) Audit

The auditor(s) may identify different issues during the audit. The error can be categorized against either one of the categories:

- Errors Affecting Compliance: Errors which directly affect the compliance process. **100 percent** weight has been assigned to errors that affect compliance.
- Other Errors: Errors which include non-compliance affecting errors. The weight of this reference error is **33 percent**.

c) Post-Audit

If the practice health falls under the category of unacceptable, then the practice will be selected for re-audit.

Sample audit findings: We have reviewed number of claims created between Jan 20, 2012 and Feb 20, 2012 and calculated compliance error rate.
Number of Days= 31 days (01-20-2012 to 02-20-2012)
Number of Claims=176
Compliance Error Rate=7. 0%

Based on the audit and error report, the practice is categorized as critical. Further categories are given in Table 3:

Table 3. Practice Compliance Health Category

Category	Non-Compliance
Normal	Less than 10%
Critical	Above 10%

3   Compliance Patterns for Healthcare Industry

Patterns are devised to control recurring problems in a certain context by proposing and implementing a specific solution to fix these types of problems. Existing identified patterns are not sufficient to handle health care regulations and standards requirements. There is a need to identify specific patterns for the United States Healthcare Industry known as compliance patterns or Billing Compliance Patterns, which are derived from applicable healthcare regulations and standards such as OIG guidelines. This approach may be adopted by third-party medical billing companies to limit fraud and errors and protect them from unnecessary government involvement. A sample pattern is given below:

Pattern 1: Correct or Prevent Use of Incorrect CPT Code or CPT Units within Billing Software

1) Name Correct the use of the CPT code and units.

2) The Problem How to enhance compliance and achieve maximum accuracy by incorporating artificial intelligence techniques within medical software to restrict incorrect use of CPT code and units entered by the billing team within the MBC medical billing software?

3) The Context
Billing is scheduled to start; Provider enrollment is not complete.
Remember that MBC's Billing Compliance Team is the billing team's friend.

4) The Forces
Billing team receives claims from Providers in scanned format. Sometimes it is difficult to read Providers' writing and the billing team enters claims without checking to confirm any doubts.
All members in a team do not have the same capabilities to handle billing.
All members need to develop and enhance existing capabilities.
Repetitive data entry activity reduces concentration.

5) The Solution
Implement a rule-base engine at the software level to systematically restrict the user from entering an incorrect claim;
Allocate tasks to billing team members based on their skill set. Billing compliance is highly dependent on the billing team
Conduct an audit at specific intervals, and keep the records in the billing compliance audit software for future reference. Based on the error rate, decide increments and bonuses.
It is better to assign challenging tasks to billing team members based on their capabilities to enhance their existing skills.

6) Rationale

The Billing Team member is the key player to ensure compliance. Different billing team members do not necessarily handle the claims in a similar manner.

Billing Team members handle claims differently. Some billing team members are good with follow-up versus handling the claims.

Some Billing Team members can communicate with Providers and Payers effectively.

7) Resulting Context

If tasks are being assigned to billing team members based on skills, the compliance with billing standard will be enhanced.

Systematic checks using rule-based engine will be more effective to minimize the potential errors.

Billing Team member's ignorance of billing can result in negligence while performing the task by making illogical assumptions.

Repeated use of this pattern can result in exhaustion of the billing team member or the amplification of exceptional capability.

4   Billing Compliance Using QMS Framework

The OIG requires medical companies to continually review and evaluate their billing processes' compliance with the third-party medical billing guideline requirements [4]. It is, therefore, vital for third-party medical billing companies to put into practice an audit framework to evaluate and improve compliance with billing standards.

4.1   Common Audit Framework

Here we consider the situation of MBC. This MBC has attained ISO 9001 certification demonstrating its ability to provide quality products and services. One of the ISO 9001 certification requirements is that MBC should conduct periodic internal audits. In this section we propose that the existing MBC quality audit framework can be tailored to comply with OIG internal monitoring and auditing requirements and thus avoid duplication and complications in the audit process. To achieve a common audit framework, inter-mapping between ISO clauses and OIG requirements was carried out by performing an in-depth comparative analysis of these two standards. After gap identification the next step is to ensure that OIG requirements are properly incorporated in the QMS so as to enhance compliance [4].

4.2   Comparative Assessment

In the section we have defined a number of operators which can be used to compare both the standards [5]. Identified operators used to compare the scope and intention of ISO and OIG standards are listed as follows:

Table 4. Comparison Standards Operators

	Title	Description
Overlap	ISO==OIG	OIG and ISO requirements are equivalent.
	ISO>OIG	ISO requirement(s) is (are) greater than OIG requirement(s) and the ISO requirement(s) include(s) OIG requirement.
	OIG>ISO	OIG requirement(s) is (are) greater than the ISO requirement(s) and OIG requirement(s) include(s) OIG requirement.
Not found	! OIG	OIG requirement does not exist in ISO 9001 standard and OIG requirement is less than the ISO requirement (ISO>(! OIG)).
	! ISO	Requirement not found in the ISO 9001 standard. ISO requirement is less than OIG requirement (OIG>(! ISO)).
Compliance Attributes	Yes	The requirement is architectural in nature and it describes the condition to which a system must conform.
(CA)	No	The requirement is not architectural in nature.

4.3   Identification, Prioritization and Comparison of Billing Compliance Requirements

The OIG has defined seven elements of requirements which we have mapped against ISO 9001 clauses as given in Table 5:

Table 5. OIG Requirements Comparison with ISO 9001 Requirements

CR#	OIG Requirement	ISO 9001 Requirement	CA?	Comparison
1	Implementing written policies, procedures, and standards of conduct	4.2.3-Control of Documents and 4.2.4-Control of Records	ARCHREQ=No	ISO==OIG
2	Designating a compliance officer and compliance committee	6.2-Human Resource	ARCHREQ=No	OIG>ISO
3	Conducting effective training and education	6.2.2-Competence, Training, and Awareness	ARCHREQ=Yes	ISO==OIG
4	Developing effective lines of communication	5.5.3-Internal Communication	ARCHREQ=Yes	ISO==OIG
5	Enforcing standards through well publicized disciplinary guidelines	6.2-Human Resource	ARCHREQ=No	OIG>ISO
6	Conducting internal monitoring and auditing	8.2.2-Internal Audit, 8.2.3-Monitoring and Measurement for Processes and 8.2.4 - Monitoring and Measurement of Products	ARCHREQ=Yes	ISO==OIG
7	Responding promptly to detected offenses and developing corrective action	8.5.2-Corrective Action 8.5.3-Preventive Action	ARCHREQ=Yes	ISO==OIG

Compliance Requirement (CR)#1: OIG requires that billing companies should realize written policies, procedures, and standards of conduct. As shown in above Table, this OIG requirement can be directly mapped to the ISO Documentation and Record requirements. We are not considering this requirement as architectural.

CR#2: OIG requires that billing companies should designate a responsible person to devise and execute policies, procedures, and standards of conducts. This can be mapped to the ISO Human Resources requirement but, in this case, the OIG requirement is greater than the ISO requirement. We are not considering this requirement as architectural.

CR#3: OIG requires that billing companies should employ an effective training program. This requirement can be mapped to the ISO Competence, Training, and Awareness requirement. We are considering this requirement as an architectural requirement based on the assumption that training is available online via an employee website.

CR#4: OIG requires that billing companies should employ effective lines of communication. This requirement can be mapped to ISO Internal Communication requirement. We are considering this requirement as an architectural requirement based on the assumption that the communication channel is available online via employee website.

CR#5: OIG requires that billing companies should enforce standards by devising and implementing disciplinary guidelines. This requirement can be mapped to the ISO Human Resources requirements. We are not considering this requirement as architectural.

CA#6: OIG requires that billing companies should conduct internal auditing and monitoring processes. This requirement can be mapped to the ISO Internal Audit and Monitoring of Processes requirements. We are considering this requirement as an architectural requirement based on the assumption that auditing and monitoring are built-in features of billing software.

CR#7: OIG requires that billing companies should promptly correct and prevent billing issues. This requirement can be mapped to the ISO Corrective & Preventive Action requirements. We are considering this requirement as an architectural requirement based on the assumption that corrective and preventive actions are implemented using billing software.

Table 6 shows the comparison summary of ISO and OIG standards. It can be concluded that 71% of OIG requirements can be directly mapped against ISO requirements. The remaining 29% can be added by the MBC to existing quality system procedures (QSP).

Table 6. Summary of Comparison of ISO 9001 and OIG Standards

	Title	**Percentage**
Overlap	ISO==OIG	71%
	ISO>OIG	0%
	OIG>ISO	29%
Not	! OIG	0%
found	! ISO	0%

Table 7 shows the OIG requirements classification according to their architectural/non-architectural nature. According to the analysis, 57 percent of OIG requirements are architectural in nature and can be directly incorporated at the system level. For the remaining 43 percent compliance can be ensured by adding the requirements to the existing QMS.

Table 7. OIG requirements Classification

	Title	**Percentage**
OIG Architectural Requirement	Yes	57%
	No	43%

5   Evaluation

MBC used this analysis and incorporated OIG guideline requirements into existing QMS processes. The next step involved an evaluation of this customized framework's effectiveness.

5.1   Audit

MBC conducted two internal ISO audits in 2010. In the first audit (IA-1), only ISO requirements were evaluated. In second audit (IA-2), OIG requirements were evaluated along with ISO requirements. Figure 8 shows the non-conformance trend in IA-1 and IA-2.

The non-conformance rate increased by 12 percent from IA-1 to IA-2. The reason is quite obvious; in IA-2, OIG requirements were also included in the scope of the internal audit.

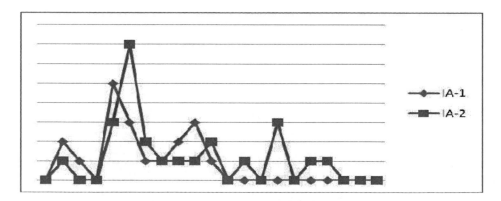

Fig. 8. ISO internal audit

5.2   Customer Satisfaction

The basic purpose of the ISO framework is to improve customer satisfaction through enhanced quality and compliant services. Four surveys were conducted to measure customer satisfaction before and after implementing the combined audit framework. It is clear from the analysis that during IA-1 (Q1 and Q2), customer satisfaction declined. One of the obvious reasons is that OIG requirements were not effectively incorporated within MBC processes.

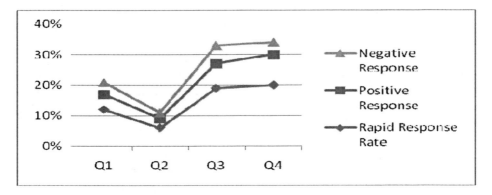

Fig. 9. Quarter-wise customer satisfaction trend

However, during IA-2 (Q3 and Q4), customer satisfaction increased tremendously. One of the major reasons for the increase in customer satisfaction is that OIG requirements were incorporated within MBC's existing QMS. Table VIII shows that customer satisfaction improved from 72 percent to 84 percent, and customer dissatisfaction was reduced from 28 percent to 16 percent.

Table 8. Customer satisfaction trend

	Start	End
Positive Response Rate	72%	84%
Negative Response Rate	28%	16%

In summary, we can conclude that customer satisfaction has increased as a result of implementing compliance patterns and the ISO and OIG combined audit approach.

6   Conclusion

OIG provides a specific billing compliance framework for medical billing companies. This paper presented a new approach for third-party medical billing companies to integrate OIG guideline requirements into an existing QMS system with minimal additional efforts. The ISO 9001:2008 guidelines might be adopted and implemented by any type of company to show its ability to provide quality services and products. This paper described the possibility of incorporating OIG requirements in the medical company's existing QMS. A billing compliance pattern was explored to enhance compliance with billing standards, e.g., OIG guideline. MBC, with a backup office in Pakistan, was used as a case study. The proposed approaches were implemented within MBC to evaluate their effectiveness. As per our analysis, the controls mentioned in the ISO 9001 guideline meet or exceed the OIG guideline for 71% of the implementation requirements. Medical companies can adopt an ISO 9001 guidelines and processes-based approach to achieve OIG compliance and improved customer satisfaction. MBC was able to achieve an increase in client satisfaction. OIG guidelines also provided a process-based model to achieve and ensure continuous compliance with its requirements. Most OIG requirements can be mapped against clauses mentioned in the ISO 9001 guidelines. The exceptions can be dealt with by devising specific procedures to address the OIG requirement and then incorporating these procedures into the existing ISO protocol.

Acknowledgment. The authors would like to acknowledge the Higher Education Commission (HEC), and National University of Computer & Emerging Sciences (FAST-NUCES) for providing required resources to complete this work. It would have been impossible to complete this effort without their continuous support.

References

1. OIG, Pub. L. 104-191, 110 Stat. 1936 (August 21, 1996)
2. Naveh, E., Marcus, A.: When Does the ISO 9000 Quality Assurance Standard Lead to Performance Improvement? Assimilation and Going Beyond. IEEE Transactions on Engineering Management 51(3), 352 (2004)
3. Gardazi, S.U., Shahid, A.A.: Billing Compliance Assurance Architecture for Healthcare Industry (BCAHI). Computer Science Journal (2010)
4. Breaux, T., Anton, A.: Analyzing Regulatory Rules for Privacy and Security Requirements. IEEE Transactions on Software Engineering (January 2008)
5. Gardazi, S.U., Shahid, A.A.: HIPAA and QMS based architectural requirements to cope with the OCR audit program, MUSIC 2012 (2012)
6. Kim, S., Kim, D.K., Lu, L., Park, S.: Quality-driven. Architecture Development Using Architectural Tactics, J. Syst. Softw. 82 (August 2009)
7. Ghanavati, S., Amyot, D., Peyton, L.: A Requirements Management Framework for Privacy Compliance. In: WER (May 2007)

Towards Personalized Medical Document Classification by Leveraging UMLS Semantic Network

Kleanthi Lakiotaki, Angelos Hliaoutakis, Serafim Koutsos,
and Euripides G.M. Petrakis

Technical Univ. of Crete (TUC), Dept. of Electronic and Computer Engineering,
Chania, Crete, Greece
klakiotaki@isc.tuc.gr, {angelos,euripides}@intelligence.tuc.gr,
skoutsos@gmail.com

Abstract. The overwhelmed amount of medical information available in the research literature, makes the use of automated information classification methods essential for both medical experts and novice users. This paper presents a method for classifying medical documents into documents for medical professionals (experts) and non-professionals (consumers), by representing them as term vectors and applying Multiple Criteria Decision Analysis (MCDA) tools to leverage this information. The results show that when medical documents are represented by terms extracted from AMTEx, a medical document indexing method, specifically designed for the automatic indexing of documents in large medical collections, such as MEDLINE, better classification performance is achieved, compared to MetaMap Transfer, the automatic mapping of biomedical documents to UMLS term concepts developed by U.S. National Library of Medicine, or the MeSH method, under which documents are indexed by human experts.

Keywords: Health Information System, User profile, MeSH, MEDLINE, AMTEx, MetaMap Transfer, Semantic Network, Recommender Systems, MCDA.

1 Introduction

In the world of medical information literature, two major categories of information seekers are mainly identified. The first, often called "healthcare consumers", or simply consumers, represent those who search in the medical document corpus to find medical information described in simple words, as opposed to "healthcare experts", that often represent medical professionals.

Over the last few decades, consumer involvement in health care has been significantly increased. At the same time, the growing health information available on the Internet offers a valuable tool to healthcare consumers. On February 2011 a research by the Pew Research Center's Internet and American Life Project and the California HealthCare Foundation (CHCF) found that 80% of Internet users look online for health information, making it the third most popular online search among all those tracked by the Pew Internet Project, following email and search engines [1].

G. Huang et al. (Eds.): HIS 2013, LNCS 7798, pp. 93–104, 2013.
© Springer-Verlag Berlin Heidelberg 2013

On the other hand, medical information systems such as MEDLINE[1] are designed to serve health care professionals (expert users in general, such as clinical doctors or medical researchers). Typically, expert users are familiar with the type and content of the medical resources (such as the National Library of Medicine - NLM dictionaries and databases) they are using and use medical terminology for their searches. A medical information system must be capable of providing dedicated, domain specific answers to experts or, simple, easy comprehend answers to novice users, respectively. A similar categorization of medical information applies in existing systems such as MedScape[2], MedlinePlus[3], Wrapin[4] and MedHunt[5] (maintained by HON, the Health on the Net Foundation, a non-profit organization aiming at providing authoritative and trustworthy information on the Web) and in other related systems such as Med-Worm[6], a medical RSS feed provider as well as a search engine built on data collected from RSS feeds. PubMed[7] of NLM is of particular interest to us. It provides free access to MEDLINE document abstracts and to articles in selected life sciences journals not included in MEDLINE.

An automatic system able to characterize medical articles as "consumer specific" or "expert specific" and thus appropriately recommend it, is valuable to both cases, by assisting consumers in managing their personal health information and experts in significantly reducing their effort on information seeking task.

In this work, we investigate on potential improvements to the problem of medical document classification by user profile (i.e., consumer users and domain experts). The high classification performance achieved in this work results from the main contributions of this paper, which are: (1) the realization and demonstration that different terms representing a medical document, contribute unequally to its classification; (2) the incorporation of Multiple Criteria Decision Analysis (MCDA), as a method for calculating the significance by which each term participates in the document classification; and (3) the representation of medical documents by term vectors extracted by the AMTEx method [2]. Evaluation results are taken on a subset of MEDLINE documents, the premier bibliographic database of the U.S. National Library of Medicine[8] (NLM). Building upon AMTEx and MetaMap Transfer[9] we show that document representations are semantically compact and more efficient, being reduced to a limited number of meaningful multi-word terms (phrases), rather than by large vectors of single-words (as it is typical in classic information systems work) part of which may be void of distinctive content semantics. Although document contents are summarized by only a few terms, these terms can be any term in the MeSH[10] with almost

---

[1] http://www.nlm.nih.gov/bsd/pmresources.html
[2] http://www.medscape.com/
[3] http://www.nlm.nih.gov/medlineplus/
[4] http://www.wrapin.org/
[5] http://www.hon.ch/HONsearch/Patients/medhunt.html
[6] http://www.medworm.com/
[7] http://www.ncbi.nlm.nih.gov/pubmed/
[8] http://www.nlm.nih.gov/
[9] http://ii.nlm.nih.gov/MMTx.shtml
[10] http://www.ncbi.nlm.nih.gov/mesh/

24,000 terms, meaning that MCDA should treat any MeSH term as a separate classification criterion which is prohibitive in practice. In this work, application of MCDA is enabled by mapping MeSH terms to their more abstract category terms in the Semantic Network[11] (SN) of UMLS[12] (Unified Medical Language System).

Related work is discussed in Section 2. Our method on document categorization by user profile is presented in Section 3. Evaluation results are presented in Section 4 followed by conclusions and issues for further research in Section 5.

2   Background and Resources

Recommender Systems (aka Recommenders or Recommendation Systems) have gained increasing popularity on the web, both in research and in industry [3]. Multi Criteria Recommender Systems have been also proved successful in this direction [4]. At the same time, in the healthcare literature, Recommender Systems are still on their infancy. In [5] for example, a health recommendation system architecture is proposed using rough sets, survival analysis approaches and rule-based expert systems. Their main goal was to recommend clinical examinations for patients or physicians from patients' self-reported data. The challenges and opportunities of merging recommender systems with personalized health education can be found in [6].

Medical document repositories, such as MEDLINE and PubMed, contain a huge amount of medical literature and are supported by NLM. Automatic extraction of useful information from these online sources remains a challenge because these documents are unstructured and indexed by human experts by assigning to each one, a number (typically 10 to 12) of terms, based on a controlled list of indexing terms, deriving from a subset of the UMLS Metathesaurus, the MeSH (Medical Subject Headings) thesaurus.

Automatic indexing and categorization of medical documents relies mainly on term extraction for the identification of discrete content indicators, namely index terms. Traditional indexing techniques ignore multi-word and compound terms, which are split into isolated single word index terms. However, compound and multi-word terms are very common in the biomedical domain [7] and are often used in indexing medical documents. Multi-word terms carry important classificatory content information, since they comprise of modifiers denoting a specialization of the more general single-word, head term. For example, the compound term "heart disease" denotes a specific type of disease.

In this work, we focus our attention on multi-word terms and AMTEx for extracting multi-word terms from medical documents. AMTEx and MetaMap Transfer have been shown to be more suitable than single-word term extraction methods not only for document indexing and retrieval, but also, for general concept description and ontology construction tasks [8]. AMTEx, in particular, has been shown to be more selective than the MetaMap Transfer (MMTx) method of NLM which maps arbitrary text to concepts in the UMLS Metathesaurus.

---

[11] http://www.nlm.nih.gov/pubs/factsheets/umlssemn.html
[12] http://www.nlm.nih.gov/research/umls/

2.1   Unified Medical Language System (UMLS)

The Unified Medical Language System (UMLS) is a source of medical knowledge developed by the U.S. NLM. UMLS consists of the Metathesaurus, the Semantic Network and the SPECIALIST lexicon. Metathesaurus is a large, multi-purpose, and multi-lingual vocabulary database. It integrates about 800,000 concepts from 50 families of vocabularies. The Semantic Network (SN) consists of 134 semantic types categorizing the Metathesaurus concepts. The purpose of the SN is to provide a consistent categorization of all concepts represented in Metathesaurus and a set of useful relationships among these concepts.

2.2   The MeSH Thesaurus

The MeSH Thesaurus (Medical Subject Headings) is a taxonomy of medical and biological terms and concepts suggested by the U.S NLM. The MeSH terms are organized in IS-A hierarchies, where more general terms, such as "chemicals and drugs", appear in higher levels than more specific terms, such as "aspirin".

2.3   WordNet

WordNet[13] is an on-line lexical reference system developed at Princeton University which attempts to model the lexical knowledge of a native speaker of English. WordNet v.2.0 (2006) contains around 127,361 terms, organized into taxonomic hierarchies. Nouns, verbs, adjectives and adverbs are grouped into synonym sets (synsets). The synsets are also organized into senses (i.e., corresponding to different meanings of the same term or concept).

2.4   MEDLINE and OSHUMED

MEDLINE database is a collection of biomedical articles. It consists of medical publications abstracts together with metadata, which is information on the organization of the data, the various data domains, and the relations between them.

The OHSUMED[14] test collection is a set of 348,566 references from MEDLINE, consisting of titles and/or abstracts from 270 medical journals over a five–year period (1987–1991). OHSUMED is commonly used in benchmark evaluations of IR applications and provides 64 queries and the relevant answer set (documents) for each query.

2.5   MetaMap Transfer Technology Transfer and AMTEx

MetaMap Transfer Technology (MMTx) uses the Metathesaurus and SPECIALIST lexicon knowledge resources during the term extraction process. This process maps arbitrary text to Metathesaurus term concepts.

---

[13]http://wordnet.princeton.edu/
[14]http://ir.ohsu.edu/ohsumed/ohsumed.html

AMTEx [2], implements the C/NC-value [9], a domain-independent method for the extraction of multi-word and nested terms. In this approach, noun phrases are initially selected by linguistic filtering. The subsequent statistical component defines the candidate noun phrase termhood by two measures: C-value and NC-value.

2.6 Multicriteria Decision Analysis

In Decision Sciences, the field of Multiple Criteria Decision Analysis (MCDA) is well established and comes into a large variety of theories, methodologies, and techniques [10]. Here, we adopt the disaggregation-aggregation approach [11], a common approach in the field of MCDA, by exploiting the UTASTAR algorithm, a representative algorithm of this approach. In abstract, the UTASTAR algorithm, considers as input a weak-order preference structure on a set of alternatives, together with the performances of the alternatives (here the medical documents) on all attributes, and returns as output a set of additive value functions based on multiple criteria, in such a way that the resulting structure would be as consistent as possible with the initial structure given by the user. This is accomplished by means of special linear programming techniques. UTASTAR's output involves the value functions associated to each criterion, approximated by linear segments, as well as the criteria significance weights (trade-offs among the criteria values).

3   Medical Document Recommendation by User Profile

We follow a three phase (*data retrieval and term extraction, data representation and modeling* and *document classification*) methodology, as described in Figure 1, to prove our assumption that certain Semantic Network sub-category terms are more important that others in medical document classification based on user profile.

In this work, we advocate that the characterization of a medical document as "for experts" or "for consumers" depends on the expert terms that the document contains, which in turn are mapped to the Semantic Network sub-category terms. The latter,

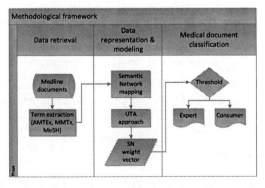

Fig. 1. The proposed methodological framework consisting of three phases (data retrieval, modeling and categorization)

represent the criteria on which the overall percentage of expert terms in a document term vector depends.

To begin, in Sections 3.1, 3.2 and 3.3, we discuss the overall framework of our proposed approach together with all the individual components.

3.1   Data Retrieval and Expert Term Extraction

As already stated, our approach relies on popular knowledge and algorithmic resources namely, the UMLS MeSH and Semantic Network for document indexing, MetaMap Transfer and the $AMTE_X$ extension for term extraction and the MEDLINE collection of biomedical articles for testing their performance.

The proposed approach for categorizing MEDLINE documents by user profile relies on the observation that MeSH terms are distinguished into i) general medical terms expressing known concepts (e.g., "pain","headache") which are easily conceived by all users, ii) domain specific terms which are used mainly by experts, iii) general - non medical terms. The more expert terms a document contains, the higher its probability to be a document for experts [12]. Moreover, since the amount of expert terms in a document is low, even for expert documents, we ignore consumer terms during the modeling process in our experiments and we represent all documents based on expert terms. Consequently, we assume mutual exclusion, meaning that any medical document that is not expert is presumably a consumer document.

We combine information from WordNet and MeSH to construct the following three term vocabularies and exploit them to characterize terms as expert or consumer:

Vocabulary of General Terms (VGT): these are terms that belong to WordNet but not to MeSH:

$$VGT = (WordNet) - (MeSH)$$

It follows that VGT contains 105,675 general (WordNet) terms.

Vocabulary of Consumer Terms (VCT): these are terms that belong to both, WordNet and MeSH:

$$VCT = (WordNet) \cap (MeSH)$$

It follows that VCT contains 7,165 consumer (MeSH) terms.

Vocabulary of Expert Terms (VET): these are MeSH terms that do not belong to WordNet:

$$VET = (MeSH) - (Wordnet)$$

It follows that VET contains 16,719 expert (MeSH) terms. Next, documents are represented by terms, extracted by applying three different approaches: AMTEx, MetaMap Transfer and the original MeSH terms.

3.2   Data Representation and Modeling

Since the Disaggregation-Aggregation approach of MCDA has mainly focused on the development of comprehensible decision models from small data sets, whose main objective is to support decision aiding through an interactive model calibration process, the total number of expert terms found in the document is too large in order to consider all the initial MeSH terms extracted from this document as criteria in the MCDA process. Furthermore, the entire set of alternatives (here medical documents) should be evaluated on the same number of criteria. For these reasons, after the term extraction process, every term originating from either AMTEx or MetaMap Transfer is mapped by the two-layered indexing structure to the UMLS Semantic Network category terms and these sub-categories are considered as criteria. MeSH terms as given by Medical Subject Headings Section staff for every document are also similarly mapped.

Only expert terms, as described in Section 3.1, count in this process and the simple term frequency measure is applied. Hence, a document in the dataset is represented by a 130-dimensional vector of expert term frequency as:

$$d_i = \{tf_1, tf_2, \ldots tf_n\}, \qquad where\ n = 1, 2, \ldots 130$$

and

$$tfi = \frac{\sum_{j=1}^{k} t_j \rightarrow t_j \in VET}{N} \tag{1}$$

where k is the number of expert terms that belong to the ith SN category and N is the total number of expert terms in d_i. For example, consider that for a document d_i five different expert MeSH terms are extracted by AMTEx, two of which belong to the sub-category "Molecular Function", one to the sub-category "Cell" and the remaining two to "Disease or Syndrome". Then, the value of sub-categories "Molecular Function" and "Disease or Syndrome" will be 2/5, while of "Cell", 1/5. Therefore, the smaller the number of a sub-category, the less this sub-category contributes to the classification of d_i as expert document. A zero value here means that no expert term from this SN sub-category was extracted. The question that arises at this point and we also try to address in this work is: *Do all sub-category of the Semantic Network contribute identically to the characterization of a document as expert? And if not, what is the significance of each sub-category for medical document classification as expert or consumer?*

MCDA methods usually assume that only a small reference set is available, since it is difficult for decision makers to express their global preferences on too many alternatives. Therefore, during the 10-fold stratified cross validation that was performed to estimate the performance of our predictive model, which was built based on the aforementioned Multicriteria Decision Analysis approach, the training set was every time randomly split into 10 segments of size n with replacement and the average values of the estimated UTASTAR parameters were calculated. By applying the so called UTASTAR algorithm in the training set, a vector of significance weights for the UMLS Semantic Network category terms is calculated, indicating the different role of each category in the characterization of a MEDLINE document as "for consumers" or "for experts".

According to the methodological requirements of the Disaggregation-Aggregation approach, a weak preference order of the alternatives (here, the medical documents), is required to apply ordinal regression. In our experiments, the probability of a document to be considered as "expert" or "consumer" is calculated as the number of expert terms that the specific document contains divided by the total number of terms extracted from the specific document. More specifically, this probability is calculated as the percentage of terms that belong to the VET. For example, a document with VET% = 0.62 has 62% probability of being a document suitable for experts. Therefore, we assume that this probability represents the global estimation of a document as expert and based on that, we transform the initial global ratings into a ranking order for the training set.

3.3   Medical Document Classification

During the last phase, the classification phase, a medical document is labeled as either expert or consumer. This choice is based upon the utility score calculated based on the final solution that corresponds to the marginal value functions (criteria weights). A linear transformation of the form $\sum_i w_i \cdot b_i$ where i denotes the SN sub-categories and b_i the value of the document under consideration for the specific SN sub-category provides the utility score for every document that belongs to the test set. The main concern that arises at this point is threshold selection. Statistical classifiers, such as a Naive Bayes classifier or Neural Network classifiers, usually calculate a score, representing the degree to which an instance is a member of a class. These scores represent probabilities, in which case standard theorems of probability can be applied. In our case however, utility scores, are not strict probabilities, in which case the only property that holds is that a higher score indicates a higher probability. Thus, in such "scoring classifiers" a threshold is necessary to produce a discrete (binary) classifier.

In diagnostic studies, the ROC curve, a plot of a test's sensitivity versus (1-specificity) for every possible threshold value, and the area under the ROC curve (AUC) are important tools in assessing the diagnostic utility of classifiers [13]. Still, finding an optimal cut-point for discriminating between binary classes is also of paramount importance.

Although the area under the ROC curve (AUC) is the most commonly used global index of diagnostic accuracy the Youden Index [14] has also frequently used by researchers. This index can be defined as $J=max_i\{Sensitivity(i) + Specificity(i) - 1\}$ and ranges between 0 and 1.

4   Evaluation

As a proof of concept we designed a series of experiments whose purpose is twofold: First, to study and compare the effectiveness of AMTEx, MetaMap Transfer and MeSH, in classifying medical documents and second, to prove that Semantic Network categories contribute differently in classifying medical documents.

Initially, documents are retrieved from a subset of the OHSUMED TREC collection consisting of 10% of OHSUMED (i.e., 34,000 documents). Both, data store and access mechanisms are implemented using Lucene. The retrieved documents were evaluated manually by users, as consumer or expert documents and this categorization is considered as the ground truth in our experiment. Subsequently, we extracted only expert documents that contain at least 2 Semantic Network category terms resulting in a subset of 237 different expert documents. To avoid any inconsistencies originating from our dataset, the same number of consumer documents is selected. Therefore, our experimental data set consists of the above 237 expert documents and 237 consumer documents and is used in all our experiments below.

As already stated, we advocate that not all the Semantic Network sub-categories contribute identically to the classification of medical documents as expert, or consumer. On the contrary, we identify different significance weights for those sub-categories. It is crucial to mention at this point that the significance weights do not follow the frequency appearance of the sub-categories in the document corpus, meaning that the most frequent sub-category is not necessarily the most important.

To evaluate classification performance we calculated several classification accuracy measures (precision, recall, F-measure, classification accuracy), in 10-fold cross-validation, as well as ROC Analysis, which altogether help us to identify the best classification performance.

Table 1 summarizes the results of best-case, worst-case and average case values of all the evaluation measures calculated on AMTEx, MetaMap Transfer and MeSH term vectors from the 10 folds. The best and worst values are assigned in bold and italics, respectively.

It is obvious from Table 1 that AMTEx outperforms all the other methods in the majority of performance measures, whilst MeSH cannot adequately separate expert and consumer documents. Moreover, it is crucial to mention here, that UTASTAR weights achieve much better evaluation scores compared to a relative equal weight scenario, by simply adding the SN category values into a utility score and apply exactly the same evaluation measures.

Table 1. Classification evaluation measures for AMTEx, MetaMap Transfer and MeSH

		AMTEx	MetaMap Transfer	MeSH
F-measure	Best -case	**0.8000**	0.3175	*0.2615*
	Worst -case	**0.5333**	0.2581	*0.1977*
	Average-case	**0.6527**	0.2814	*0.2261*
Precision	Best -case	**0.7097**	0.1942	*0.1604*
	Worst -case	**0.3636**	0.1481	*0.1149*
	Average-case	**0.5085**	0.1660	*0.1313*
Recall	Best -case	**1.0000**	**1.0000**	**1.0000**
	Worst -case	**0.9130**	0.8696	*0.6957*
	Average-case	**0.9449**	0.9322	*0.8391*
AUC	Best -case	**0.9732**	0.7293	*0.6578*
	Worst -case	**0.9528**	0.6407	*0.5359*
	Average-case	**0.9629**	0.6882	*0.6047*
Accuracy	Best -case	**0.9580**	0.6705	*0.6336*
	Worst -case	**0.8397**	0.4733	*0.3359*
	Average-case	**0.9037**	0.5656	*0.4743*

Additionally, Figure 2 illustrates the ROC curves for the best precision values in all cases allowing us to compare the classification performance of our model built based on AMTEx terms, MetaMap Transfer terms and MeSH terms.

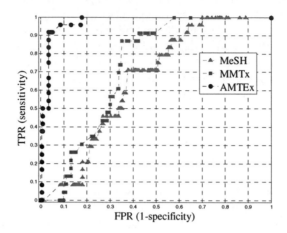

Fig. 2. ROC curves for AMTEx, MetaMap Transfer and MeSH approaches

Clearly, AMTEx shows the best performance than MetaMap Transfer or MeSH.

We also studied the classification performance of the three medical document representation methods by applying Decision Tree analysis (DTA), as an alternative of Multiple Criteria Decision Analysis. Decision trees are a simple, but powerful form of multiple attribute analysis. Decision trees are produced by algorithms that identify various ways of splitting a data set into branch-like segments. Here, we consider two segments, experts and consumers, as the dependent class variables and the Semantic Network sub-categories as the independent variables. We performed 10 fold cross-validation on the J48 classifier with default parameter values (confidence threshold for pruning at 0.25) for inducing classification trees. The J48-algorithm is a Java re-implementation of the C4.5-algorithm and is a part of the machine learning package WEKA[15]. Table 2 shows the results of Decision Tree Analysis on the same document collection.

Table 2. Decision Tree Analysis results

	Accuracy (%)	Precision (%)	Recall (%)	F-Measure (%)	AUC
AMTEx	83.75	87.3	83.8	80.8	0.812
MetaMap Transfer	78.9	79.5	78.9	78.8	0.807
MeSH	65.82	66.0	65.8	65.7	0.674

---

[15]http://www.cs.waikato.ac.nz/ml/weka/

Clearly, the Decision Tree analysis results, also confirm that AMTEx is the most appropriate method for term extraction and representation for the MEDLINE documents that we studied. This result becomes more apparent in the MCDA experiments, probably due to the fact that AMTEx terms seem to be more representative for the classification of medical documents, than MetaMap Transfer or MeSH terms assigned by humans, leading thus to a better identification of the significance of SN subcategories in this classification. Moreover, AMTEx achieves better results in most evaluation measures when MCDA is applied compared to DTA.

5   Conclusions

This work brings together ideas from document information management and Recommender Systems and shows how these ideas can applied for effectively classifying medical documents by user profile. We investigated the problem of automatic categorization of medical information on two common types of users (consumers and experts). Medical documents were represented by term vectors extracted from three different approaches (AMTEx, MetaMap Transfer and MeSH). Based on our experiments, we conclude that when documents are represented by AMTEx, a medical document indexing method, specifically designed for the automatic indexing of documents in large medical collections, such as MEDLINE, the categorization performance is significantly increased compared to when the same documents are represented by MetaMap Transfer, the automatic mapping of biomedical documents to UMLS term concepts developed by U.S. National Library of Medicine, or the MeSH method, under which documents are indexed by human experts, based on a controlled list of indexing terms, deriving from a subset of the UMLS Metathesaurus. We also proved that the UMLS Semantic Network sub-category terms can act as criteria for the categorization of a medical documents, however their performance play an important role in their classification ability. Moreover, our experiments show that Multiple Criteria Decision Analysis, as a method for identifying the significance of Semantic Network sub-category terms in their classification ability, achieves better results compared Decision Tree Analysis, when documents are represented by AMTEx terms.

Future work involves extending our studies in discovering mechanisms for classifying medical documents into several thematic categories. For example, our intention is to be able to recommend medical documents to expert users for "breast cancer".

Acknowledgments. This research leading to these results has received funding from the European Community's Seventh Framework Program (FP7/2007-2013) under grant agreement No 296170 (Project PortDial) and grant agreement No 248801 (Project RT3S)".

References

1. Fox, S.: 80% of internet users look for health information online (2011)
2. Hliaoutakis, A., Zervanou, K., Petrakis, E.G.M.: The AMTEx approach in the medical document indexing and retrieval application. Data & Knowledge Engineering 68, 380–392 (2009)
3. Ricci, F., Rokach, L., Shapira, B., Kantor, P.B.: Recommender Systems Handbook. Springer, US (2011)
4. Lakiotaki, K., Matsatsinis, N., Tsoukiàs, A.: Multicriteria User Modeling in Recommender Systems. IEEE Intelligent Systems 26, 64–76 (2011)
5. Pattaraintakorn, P., Zaverucha, G.M., Cercone, N.: Web Based Health Recommender System Using Rough Sets, Survival Analysis and Rule-Based Expert Systems. In: An, A., Stefanowski, J., Ramanna, S., Butz, C.J., Pedrycz, W., Wang, G. (eds.) RSFDGrC 2007. LNCS (LNAI), vol. 4482, pp. 491–499. Springer, Heidelberg (2007)
6. Luis, F.-L., Randi, K., Lars, V.: Challenges and opportunities of using recommender systems for personalized health education. In: Medical Informatics in a United and Healthy Europe, pp. 903–907. European Federation for Medical Informatics (2009)
7. Maynard, D., Ananiadou, S.: TRUCKS: A model for automatic multi-word term recognition. Journal of Natural Language Processing 8, 101–125 (2000)
8. Divita, G., Tse, T., Roth, L.: Failure analysis of MetaMap Transfer (MMTx). Studies in Health Technology and Informatics 107, 763–767 (2004)
9. Frantzi, K., Ananiadou, S., Mima, H.: Automatic recognition of multi-word terms: the C-value / NC-value method. International Journal on Digital Libraries 3, 115–130 (2000)
10. Figueira, J., Greco, S., Ehrgott, M.: Multiple criteria decision analysis: state of the art surveys. Springer, London (2005)
11. Siskos, Y., Grigoroudis, E.: UTA methods. In: Figueira, J., Greco, S., Ehrgott, M. (eds.) Multiple Criteria Decision Analysis: State of the Art Surveys, pp. 297–344 (2005)
12. Petrakis, E.G.M., Hliaoutakis, A.: Automatic Document Categorisation by User Profile in Medline, pp. 1–10 (2011)
13. Fawcett, T.: ROC Graphs: Notes and Practical Considerations for Data Mining Researchers, HP Laboratories Palo Alto (2003)
14. Fluss, R., Faraggi, D., Reiser, B.: Estimation of the Youden Index and its associated cutoff point. Biometrical Journal 47, 458–472 (2005)

Segmentation of Retinal Blood Vessels Using Gaussian Mixture Models and Expectation Maximisation

Djibril Kaba[1,*], Ana G. Salazar-Gonzalez[1], Yongmin Li[1], Xiaohui Liu[1], and Ahmed Serag[2]

[1] Department of Information Systems, Computing and Mathematics Brunel University, United Kingdom
[2] Diagnostic Imaging and Radiology, Children's National Medical Center, Washington, DC, USA

Abstract. In this paper, we present an automated method to segment blood vessels in fundus retinal images. The method could be used to support a non-intrusive diagnosis in modern ophthalmology for early detection of retinal diseases, treatment evaluation or clinical study. Our method combines the bias correction to correct the intensity inhomogeneity of the retinal image, and a matched filter to enhance the appearance of the blood vessels. The blood vessels are then extracted from the matched filter response image using the Expectation Maximisation algorithm. The method is tested on fundus retinal images of STARE dataset and the experimental results are compared with some recently published methods of retinal blood vessels segmentation. The experimental results show that our method achieved the best overall performance and it is comparable to the performance of human experts.

Keywords: Retinal image, vessel segmentation, matched filter, bias correction, expectation maximisation.

1 Introduction

Automated segmentation of retinal structures allows ophthalmologist and eye care specialists to perform mass vision screening exams for early detection of retinal diseases and treatment evaluation. This non-intrusive diagnosis in modern ophthalmology could prevent and reduce blindness and many cardiovascular diseases around the world. An accurate segmentation of retinal blood vessels (vessel diameter, colour and tortuosity) plays an important role in detecting and treating symptoms of the retinal abnormalities such haemorrhages, vein occlusion, neo-vascularisation. However, the intensity inhomogeneity and the poor contrast of the retinal images cause a significant degradation of the performance

---

* The authors would like to thank the Department of Information Systems, Computing and Mathematics, Brunel University for financial support.

G. Huang et al. (Eds.): HIS 2013, LNCS 7798, pp. 105–112, 2013.

of automated blood vessels segmentation techniques. The intensity inhomogeneity of the fundus retinal image is generally attributed to the acquisition of the image under different conditions of illumination.

Previous methods of blood vessels segmentation can be classified into two categories: (1) pixels processing based methods, and (2) tracking-based methods [1]. Pixel processing based methods use filters to enhance the appearance of the blood vessels in the image, then processing techniques such as thinning or branching are applied to classify pixel as either belonging to vessels or not. Hoover et al. [2] proposed a framework to extract blood vessel from retinal images using a set of twelve directional kernels to enhance the vessels before applying threshold-probing technique for segmentation. Mendoca et al. [3] presented a method to extract a vessel centreline then filled it using the global intensity characteristics of the image and the local vessel width information. The drawback of these techniques is that some proprieties of blood vessels can only be applied in the segmentation process after a low level preprocessing and they generally output poor segmentation results on disease retinal images.

Tracking-based methods use a Gaussian function to characterise a vessel profile model, which estimates the location of vessel points use in vessel tracing. Xu et al. [4] combined the adaptive local thresholding method and the tracking growth technique to segment retinal blood vessels. Salazar et al. [5] used an adaptive histogram equalisation and the distance transform algorithm to enhance the vessels appearance, then applied the graph cut technique to segment vessels. The limitations of these methods is that the multiple branches models are not applicable and they do not perform well on disease retinal images.

2   Methods

In this paper, we present a new automated method to extract blood vessels in retinal funds images. The proposed method combines two pre-processing techniques. Our approach takes as a first step the correction of the intensity inhomogeneity of the retinal image using a bias correction algorithm[6] , then the appearance of blood vessels are enhanced with a matched filter response. The Expectation Maximisation (EM) algorithm is used to extract the vascular tree from the matched filter response image. Finally, a length filter is applied on the EM algorithm output image to eliminate all the non-vessels pixels. An overview diagram and experiment results of the steps used in our method is given in Fig. 1 and Fig. 2 respectively.

2.1   Bias Correction

One of the major issues associated with fundus retinal images is the intensity inhomogeneity across the images, which causes a significant degradation to the performance of automated blood vessels segmentation techniques.The intensity inhomogeneity of the fundus retinal image is generally attributed to the acquisition of the image under different conditions of illumination.

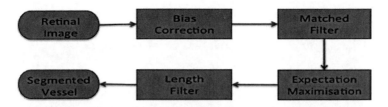

Fig. 1. Vessel segmentation algorithm

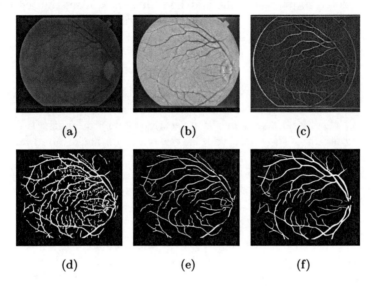

Fig. 2. (a) Retinal image with intensity inhomogeneity. (b) Bias corrected image. (c) Matched filter response image. (d) EM image. (e) Length filter image. (f) Ground truth image.

In order to overcome such a problem, we use the N4 algorithm of bias correction presented in [6] which is a modified version of the originally bias correction proposed N3 algorithm [7] that includes a modified iterative update within a multi-resolution framework. Let consider a noise free two dimensional medical image v, which can be defined as:

$$\hat{v}(i,j) = \hat{I}(i,j) + \hat{B}(i,j) \tag{1}$$

where I is the uncorrupted image, B is the bias field and $\hat{v} = \log v$, $\hat{I} = \log I$ and $\hat{B} = \log B$. To define the uncorrupted image, the following iterative equation is derived at the nth iteration as:

$$\hat{I}(i,j)^n = \hat{I}(i,j)^{n-1} - S^*\{\hat{I}(i,j)^{n-1} - E[\hat{I}(i,j) \mid \hat{I}(i,j)^{n-1}]\} \tag{2}$$

where $\hat{I}(i,j)^0 = \hat{v}(i,j)$, the initial bias field estimate and $\hat{B}(i,j)_e^0$ is equal to zero. $S^*\{.\}$ and $E[\hat{I}(i,j) \mid \hat{I}(i,j)^{n-1}]$ are the B-spline approximator and the expected value of the true image given a current estimated of the corrected image respectively. The expression of the expected value of the true image is derived in [7]. Figure. 2 (b) shows the bias corrected image.

2.2   Matched Filter for Blood Vessels

In order to enhance the appearance of the blood vessels in the retinal image, we have implemented a two-dimensional matched filter. This technique adapted the method presented in [8], where the matched filter maximises the response over blood vessels and minimises it in the image background. The intensity profile of vessels is approximated by a Gaussian function as:

$$f(x, y) = A\{1 - k \exp(-d^2 \setminus 2\sigma^2)\} \tag{3}$$

where A represents the gray-level intensity of local background, d is the perpendicular distance between given point (i, j) and the straight line passing through the centre of the blood vessel in the direction along its length, k defines the measure of contrast of the blood vessels relative to its background and σ is the spread of the intensity profile. To design an optimal filter, the filter must have the same shape as the intensity profile. Thus from the intensity profile, the weight coefficients of the two-dimensional kernel are given by

$$K(x, y) = -\exp(-x^2 \setminus 2\sigma^2) \tag{4}$$

where $|y| \leq L \setminus 2$, L denotes the length of the segment for which the blood vessel is assumed to have a fixed direction. Since blood vessels may appear in any direction, a set of twelve 16×15 pixels kernels is designed and rotated for all possible orientations of the blood vessels. Thus, the kernels are applied at every pixel in the image and only the maximum value of the matched filter response is retained. A pixel is labelled as blood vessel if this maximum value is above a certain threshold. More details for computing the weight coefficients of the two-dimensional kernel can be found in [8]. Figure. 2(c) shows the matched filter response image.

2.3   Expectation Maximisation

To extract the blood vessels from the fundus retinal image, we implemented the Expectation Maximisation (EM) algorithm adapted from [9] and applied it to a Gaussian mixture distribution of the pixel intensities. The EM performs the segmentation by classifying vessel's pixels in one class (foreground) and non-vessel's pixels in the other (background). The EM output is obtained by iteratively performing two steps: E-step, which computes the expected value of the likelihood function (pixel class membership function) with respect to the unknown variables, under the expected parameters of a Gaussian mixture model and M-step which maximises the likelihood function defined in the E-step until convergence [10]. The EM algorithm takes the value of a pixel's intensity as a random variable. Like any random variables, the pixel intensities of an image have a probability.

To derive the probability model, we assume a Gaussian mixture distribution as [11]:

$$P(x \mid \mu_i, \sigma_i^2) = \frac{1}{\sigma\sqrt{2\pi}} \exp \frac{-(x - \mu_i)^2}{2\sigma_i^2} \tag{5}$$

where (μ_i) is the mean vectors and (σ_i^2) is the variance of class i. Our aim is to estimate and maximise the parameters of the Gaussian mixture model. The likelihood estimates function (LE) of parameters in the Gaussian mixture model is expressed as:

$$L(\theta) = \sum_{N=1}^{n} \log \sum_{i=1}^{k} p_i P(x_N \mid \mu_i, \sigma_i^2) \tag{6}$$

where p_i is the mixing proportion (the probability of a pixel belonging to a class i, where $i = 1, 2, ...k$). The values of the parameters which maximise (6) are known as maximum likelihood estimators (MLE). The MLE is defined as:

$$\mu_i = \frac{\sum_{N=1}^{n} P(i \mid x_N)}{P(i \mid x_N)} \tag{7}$$

$$\sigma_i^2 = \frac{\sum_{N=1}^{n} P(i \mid x_N)(x_N - \mu)^2}{P(i \mid x_N)} \tag{8}$$

where μ_i, σi^2 and p_i are the maximum likelihood of the mean vector, the variance and the mean membership probability (expected value) receptively. More details of MLE are given in [9].

2.4 Length Filter

Figure 2 (d), shows some misclassified pixels in the EM output image. The length filtering model is designed to eliminate all the non-vessels pixels in the EM algorithm output image. We adapt the length filtering used in [12], which discard all the groups of pixel with pixel number less than a certain number of pixels. The approach uses connected pixels labelling model, in which each individual object in the image is defined as connect regions. The approach starts by identifying all the connected regions, then discard all the connected objects less than a certain number of pixels using an eight-connected neighbourhood of all surrounding pixels. Finally a label propagation is used and all connected components larger than a certain number of pixels are labeled as blood vessels. This approach reduces significantly the false positive, the output of the length filtering is shown in Fig. 2 (e).

3 Experimental Results

The method presented in this paper was evaluated on publicly available retinal image dataset STARE presented by Hoover et. al. [2]. The STARE dataset contains 20 fundus colour retinal images captured by a Topcon TRV-50 fundus camera at 35 degree field of view (FOV) and the size of the image is 700x605 pixels. We calculated the mask of the image for this dataset using a simple threshold technique for each colour channel. The dataset included images with retinal diseases selected by Hoover et al. The dataset provides two sets of hand labelled images segmented by two human experts. The first expert labelled fewer vessel pixels than the second one. We adapt the first expert hand labelled as the ground truth.

To facilitate the performance comparison between our method and other retinal blood vessels segmentation methods, we have considered the accuracy rate as performance measure. The following parameters are used in the experiment to measure the performance of the segmentation: True Positive Rate (TPR), which is defined by dividing the number of true positive by the total number of blood vessels in the ground truth image, False Positive Rate (FPR) is defined by dividing the number of false positive with the total number of non blood vessels in the ground truth image, and the accuracy (ACC) is computed as the sum of true positives and true negatives over the total number of pixels in an given image. It is worth mentioning that a perfect segmentation would have a FPR of 0 and a TPR of 1. We also tested the performance of our method on retinal disease images provided by STARE dataset, which generally cause a significant degradation to the performance of automated blood vessels segmentation techniques. The experiment values of different retinal blood vessels segmentation methods on the STARE dataset is shown in Tables 1 and 2. All the methods used the first expert hand labelled images as performance reference.

The experimental results in Tables 1 show that our method with an accuracy of 0.9450 performed better than the methods presented by Mendoca et al. [3] and Hoover et al [2] and its only marginally inferior to the hand segmented images of

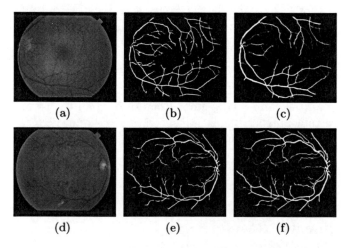

Fig. 3. (a) Healthy retinal image. (b) Final vessel segmentation. (c) Ground truth. (d) disease image. (e) Final vessel segmentation. (f) Ground truth.

Table 1. Performance of vessel segmentation methods on STARE dataset (disease and Healthy retinal images)

Method	TPR	FPR	Accuracy
2nd human observer [3]	0.8949	0.0610	0.9354
Mendonca [3]	0.6996	0.0270	0.9440
Hoover[2]	0.6751	0.0433	0.9267
EM segmentation	**0.6645**	**0.0216**	**0.9450**

Table 2. Performance of vessel segmentation methods on STARE dataset (disease retinal images only)

Method	TPR	FPR	Accuracy
2nd human observer [3]	0.8252	0.0456	0.9425
Mendonca[3]	0.6733	0.0331	0.9388
Hoover[2]	0.6736	0.0528	0.9211
EM segmentation	**0.6520**	**0.0255**	**0.9411**

the second human expert. Tables 2 shows that our method with an accuracy of 0.941 on disease retinal images, outperforms the methods proposed by Mendoca et al. [3] and Hoover et al [2] and it is comparable to the performance of human experts.

4   Conclusions and Discussion

We have presented in this paper an new approach to blood vessels segmentation by integrating the bias correction, the matched filtering and the EM segmentation method. We have evaluated our method against other retinal blood vessels segmentation methods on STARE dataset. The experimental results presented in Tables 1 and 2 show that the proposed approach achieved the best overall performance.

Our method has an advantage over tracking-based methods because it applies a two-dimensional matched filter on retinal images to enhance vessel appearance and allows multiple branches models. Also our method achieves better results over pixel processing based methods as it corrects the intensity inhomogeneities across retinal images to improve the segmentation of the blood vessels. This technique also minimises the segmentation of the optic disc boundary and the lesions in the retinal images.

However, we are aware that the improvement in term of accuracy is marginal over other methods we compared against. Also, the size of the dataset is relatively small although it is the best we have at the moment for this specific problem. Evaluating our method on a larger dataset will be part of our future work.

References

1. Fritzsche, K., Can, A., Shen, H., Tsai, C., Turner, J., Stewart, C., Roysam, B.: Automated model based segmentation, tracing and analysis of retinal vasculature from digital fundus images (2003)
2. Hoover, A., Kouznetsova, V., Goldbaum, M.: Locating blood vessels in retinal images by piecewise threshold probing of a matched filter response. IEEE Transactions on Medical Imaging 19(3), 203–210 (2002)
3. Mendonça, A.M., Campilho, A.: Segmentation of retinal blood vessels by combining the detection of centerlines and morphological reconstruction. IEEE Trans. Med. Imaging 25, 1200–1213 (2006)
4. Xu, L., Luo, S.: A novel method for blood vessel detection from retinal images. BioMedical Engineering OnLine 9(1), 1–10 (2010)
5. Salazar-Gonzalez, A., Li, Y., Kaba, D.: Mrf reconstruction of retinal images for the optic disc segmentation (2012)
6. Tustison, N., Avants, B., Cook, P., Zheng, Y., Egan, A., Yushkevich, P., Gee, J.: N4itk: Improved n3 bias correction. IEEE Transactions on Medical Imaging 29(6), 1310–1320 (2010)
7. Sled, J., Zijdenbos, A., Evans, A.: A nonparametric method for automatic correction of intensity nonuniformity in mri data. IEEE Transactions on Medical Imaging 17(1), 87–97 (1998)
8. Chaudhuri, S., Chatterjee, S., Katz, N., Nelson, M., Goldbaum, M.: Detection of blood vessels in retinal images using two-dimensional matched filters. IEEE Transactions on Medical Imaging 8(3), 263–269 (1989)
9. Dempster, A., Laird, N., Rubin, D.: Maximum likelihood from incomplete data via the em algorithm. Journal of the Royal Statistical Society. Series B (Methodological), 1–38 (1977)
10. Clarke, L., Velthuizen, R., Camacho, M., Heine, J., Vaidyanathan, M., Hall, L., Thatcher, R., Silbiger, M.: Mri segmentation: Methods and applications. Magnetic Resonance Imaging 13(3), 343–368 (1995)
11. Pyun, K., Lim, J., Won, C., Gray, R.: Image segmentation using hidden markov gauss mixture models. IEEE Transactions on Image Processing 16(7), 1902–1911 (2007)
12. Chanwimaluang, T., Fan, G.: An efficient blood vessel detection algorithm for retinal images using local entropy thresholding. In: Proceedings of the 2003 International Symposium on Circuits and Systems, ISCAS 2003, vol. 5, pp. V–21. IEEE, Los Alamitos (2003)

Modeling and Query Language for Hospitals

Janis Barzdins[1], Juris Barzdins[1,2], Edgars Rencis[1], and Agris Sostaks[1]

[1] Institute of Mathematics and Computer Science, University of Latvia, Riga, Latvia
{janis.barzdins,edgars.rencis,agris.sostaks}@lumii.lv
[2] Faculty of Medicine, University of Latvia, Riga, Latvia
juris.barzdins@lu.lv

Abstract. So far the traditional process modeling languages have found a limited use in the hospital settings. One of the reasons behind this delay has been the lack of clear definition of the sequence of activities that are carried out in the hospital. We propose a new modeling language (as a profile of UML Class diagrams) that captures all the useful features from various UML diagrams and can be used in modeling of the hospitals. Based on the modeling language, we have developed an easy-to-perceive graphical query language, which allows the physicians to retrieve directly from the various hospital databases information they need to better understand the flow of clinical processes.

Keywords: Hospital, Modeling, Modeling languages, Query languages.

1   Introduction

In 2002, the management professor and renowned author Peter Drucker stated in his book "Managing in the Next Society", that "health care is the most difficult, chaotic, and complex industry to manage today", and that the hospital is "altogether the most complex human organization ever devised". Since then, hospitals have made advances in implementation of the promising health information technologies (HIT) in hopes to achieve major healthcare cost savings, reduce medical errors and improve health outcomes [2]. Unquestionable and measurable has been the positive impact of the HIT on patient safety, quality and continuity of care, and the patient empowerment [3.4]. So far however there has been little evidence that implementation of the HIT is leading to health care cost savings [4]. One of the reasons for this lack of impact by the HIT likely lies in the complexity of the business process ownership in the hospitals.

While both the management and support processes are directly controlled by the hospital management, the main operational clinical processes which constitute the core value production of the business has generally been owned by the doctors. Since medical professionals and not the managers carry the ultimate responsibility for the patient's outcomes, the management has a limited control over the doctors' individual bedside decisions. Therefore, a more profound involvement of the doctors in transforming the processes within their health care organizations has been widely regarded as a factor that is critical for their success [5-8]. This is particularly true considering the fact that up to 85% of all the spending in health care is directly or indirectly controlled by the medical professionals [9].

G. Huang et al. (Eds.): HIS 2013, LNCS 7798, pp. 113–124, 2013.

In contrast to the professional managers who have received an appropriate training and control the administrative resources (e.g., specially dedicated business analysts for extracting process knowledge from the increasing amount of digitally stored data), doctors so far have benefitted to a much lesser degree from these advances in HIT as a tool for better understanding of the patterns and systemic consequences of the clinical decisions they make. The goal of our research is to develop a business model-based method for hospital use which would allow doctors to retrieve directly the ad-hoc information from various hospital databases which is needed in building their process-oriented knowledge for their managerial roles.

For better understanding, we broke down the task of achieving this goal into two steps. First, we developed a new domain-specific language for hospital modeling which allows doctors and managers to visualize the hospital processes (see Section 3). Subsequently, based on this modeling language, we developed an easy-to-perceive graphical query language which permits retrieving specific information needed for the analysis of a particular clinical process. The query language described in Section 4 is considered to be the basic added value of this paper. An evaluation of our approach is given in Conclusions.

2   Related Work

In recent years business processes in hospitals have been studied for the applicability of modeling methods used in other industries. For example, there are published reports of successful usage of BPMN for describing the clinical process for strictly selected group of patients with a specific diagnosis in oncology [10] and the process in selected department for pathology investigations [11]. However, there are also reports suggesting that application of BPMN is difficult in the specific domain of health care, since the nature of health care processes in a multidisciplinary hospital is inherently complex [12], and that has been the basis also for the domain-specific modeling in testing [13].

There are works on querying the descriptions of the business processes without the underlying data, e.g., work of Beeri et.al. [14], where the visual query language BP-QL has been introduced, and the BPMN-Q language by Awad [15].

Beeri et.al. [16] went a step further by introducing the BP-Mon – a query language for monitoring business processes, which allows the users defining the monitoring tasks and retrieving their associated reports visually. Although the language is simple enough for IT specialists, it is hardly useable by doctors in the hospitals. For example, the specification of reports (retrieved data) requires knowledge about XML.

Beheshti et.al. [17] introduced a process mining and querying methodology, where data are acquired also from the information system. These data are called *Event Logs* and are grouped into *folder nodes* – a similar concept to *slices* presented in this paper. However, the query language is itself based on SPARQL making it impractical for a broader use by the hospital staff.

3   Hospital Modeling Language MEDMOD

The most-widely known general purpose modeling language UML offers at least three different types of languages, whose elements can be used in hospital modeling.

The first one – UML Activity diagrams (and also the BPMN diagrams) – describes the sequence of activities to be performed. However, this kind of language cannot be directly used in the hospital domain because of the large degree of variations of the order how doctor execute various treatment procedures. The sequence of the activities can only be partially defined here. On the other hand, certain procedures have established protocols and well-defined sequence of activities, e.g. registration of the patient, or anesthesia used in performing certain types of procedures, and these are suitable for analysis using the UML activity diagram. Therefore, while the activity diagrams can be used in describing some aspects of the hospital operation, they are not applicable for the entire process.

The second type of diagrams are the UML Class diagrams or ontologies (they largely differ in only one aspect – the former uses closed-world semantics, while the latter exploits the benefits of the open-world semantics, e.g., see [18]). This type of diagrams is very convenient for concept modeling, but is not oriented towards modeling the activities. UML allows however perceiving an activity, e.g., X-ray investigation for a patient, as a class. Instances of this class would then be defined as certain X-ray investigations used for specific patients. Further in this paper we will make an essential usage of this type of classes.

The third type of diagrams is UML Use-case diagrams, which combine the elements of the class and activity diagrams. They describe activities called use-cases. It is also stated that use-cases can be perceived as classes, whose instances are concrete executions of these activities. A very useful aspect of the use-case diagrams is their capability for interaction between the use-cases with extending the activity by calling another activity. In other words, the extension point mechanism in the use-case diagrams makes it possible to describe specific control flows having a guard condition (the extension point name), which are executed during the current activity instead of waiting for the activity to complete. This feature isn't present neither in UML activity diagrams, nor in BPMN, but is very pertinent in the case of hospital modeling. At the same time there are no ordinary control flows in use-case diagrams, because use-case diagrams are a priori dedicated to describing a higher-level functionality.

This all led us to think that a special domain-specific modeling language is needed for hospital modeling, which would borrow the most useful features from class, activity and use-case diagrams. We have developed such a language called the MEDMOD. Formally, we can define the MEDMOD as a profile on UML Class diagrams as can be seen in Fig. 1 (OCL constraints defining MEDMOD more precisely are omitted here).

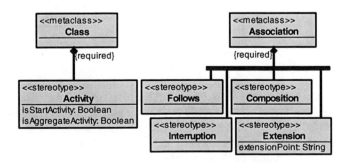

Fig. 1. The UML profile defining the MEDMOD language

We are however describing the language on examples (see Fig. 2) for its easier perception by the domain experts (doctors and managers).

Let us now describe the elements of MEDMOD in more detail (see Fig. 2).

Activity. Activity is the central element of the MEDMOD language and denotes a task in time having a start and end moments. Semantically it is related to the Action element of UML Activity diagrams. Examples of Activity are seen in Fig. 2 depicted as yellow boxes with rounded corners.

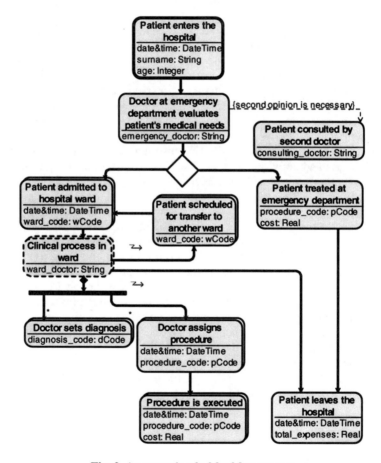

Fig. 2. An example of a MEDMOD process

From the linguistic point of view, we divide Activities in three different categories based on how the Activity name is formed. The first type of Activities is the most common one and conforms to the simple present linguistic form – "Doctor sets diagnosis". The second type of Activities is formed in passive voice and used in cases, when there can be different consequences to some previous Activity leading to execute one of different outgoing flows from it – "Patient admitted to hospital ward". The third type of Activities refers to a greater process with some given name, which then serves as a name for the Activity – "Clinical process in ward". These naming conventions are, however, only guidelines for users creating and reading the MEDMOD diagrams and they are provided for better comprehension of the process. The visual appearance of Activity does not depend on its linguistic type.

Activities can also have attributes of five primitive data types – Integer, Real, String, Boolean and DateTime. These attributes can be specified for every concrete Activity at diagram creation, and different values can be assigned to these attributes of concrete instances of the Activity at run-time. Since there are very detailed codificators in the medical world for coding every procedure, diagnosis or other attributes (see Health Level Seven International, the global authority on standards for interoperability of health information technology [19]), we also allow using enumerations as data types. For instance, the Activity "Doctor assigns procedure" has an attribute "procedure_code", whose values come from the enumeration "pCode" (see Fig. 2).

Exactly one of the Activities of every MEDMOD diagram is denoted as the Master Activity meaning that the execution of the diagram starts with this Activity (there can be no ingoing arrows to this Activity). Master Activity has a slightly different visualization – a bolder frame (see Activity "Patient enters the hospital" in Fig. 2).

Follows. This type of oriented relation can be established between two Activities A and B meaning that Activity B can only start after Activity A has ended (the same semantics as the Control Flow of UML Activity diagrams). It is allowed for several Activities to follow the same Activity – the XOR semantics is implied in this case meaning that only one of those outgoing flows can be executed. We denote this situation by introducing a new diamond-shaped graphical element seen in Fig. 2. It is also allowed to have several ingoing flows into an Activity implying the OR semantics, i.e., the following Activity can start executing when at least one flow has executed, and several instances of that Activity can arise, if several incoming flows executes at different times. It is, however, not allowed to introduce several parallel outgoing flows from the same Activity. We substitute the parallel branching of UML Activities with a more general feature, the composition, by introducing so called Aggregate Activities and their parts – Component Activities that can be executed simultaneously.

Composition. A composition between two Activities can be established, if one Activity (called the Aggregate) semantically consists of one or more other Activities (called the Components). It has an analogy with the relation "includes" of UML Use-case diagrams. We have borrowed the notation for the Aggregate part of Composition (the filled diamond) from the UML Class diagrams. Also, a composition fork graphical element can be introduced to collect the Components of the same Aggregate Activity (seen in Fig. 2). For instance, Activity "Clinical process in ward" consists of two types of Activities – "Doctor assigns procedure" and "Doctor sets diagnosis" (notice a slightly different visualization – a dashed frame – for Aggregate Activities in Fig. 2). Each Component Activity can appear several times within the Aggregate, therefore we also allow cardinalities to be attached to the Component end of a Composition (the default cardinality is 1).

Interruption. An interesting phenomenon relates the composition – what is the semantics of a Follows-type relation going out from the Aggregate Activity? It was stated before that the Follows flow can execute when the Activity A has ended. But the

Aggregate Activity can actually never end, if it has at least one Component having a cardinality, e.g., * (many). In this case the Aggregate is constantly waiting for new and new Component instances to born, and only some force from outside can decide, when to stop the waiting process. We must therefore introduce a new type of control flow – an Interruption – stating that if there is an outgoing Interruption flow from the Aggregate Activity A to some Activity B, it means that the Activity A is suspended, when the flow is executed (i.e., when the Activity B needs to be started) meaning that it can no more create new Component instances (already created Component instances continues to execute normally). For instance, in Fig. 2 the Activity "Clinical process in ward" is suspended when the doctor decides to either transfer the patient to another ward, or to discharge the patient. The Interruption flow is adorned with a jagged "lightning bolt" arrow. Simple Activities can also be interrupted in similar manner.

Extension. Extension is an oriented relation between two Activities A and B meaning that Activity B can be called at some time during the execution of Activity A. This feature allowing us to extend the Activity is also borrowed from UML Use-case diagrams. The call is triggered, when some predefined condition occurs. The condition is described as an Extension point name and attached to the Extension. For instance, a doctor can decide that a "second opinion is necessary" (the Extension point name) during the evaluation of patient's medical needs. In that case another Activity "Patient consulted by second doctor" is called (see Fig. 2).

Using the four abovementioned elements, one can define a MEDMOD process serving at least two purposes: 1) the visualization of hospital processes can help doctors and management of the hospital in performing their daily tasks better; 2) one can use the graphical process in order to perform queries on their underlying real data. This is one of the added values of this paper. To achieve the second part of the goal stated in the Introduction, we must first introduce a new concept of a slice being exploited in the next section. If we look at the MEDMOD diagram from the process point of view, we can notice that every instance of the Master Activity define a separate transaction consisting of those instances of Activities that can be reached from the instance of the Master Activity (these are called run-time instances in the process modeling world). We call the set of all run-time instances within a transaction a slice. The basic assumption we make here is that no two slices can ever share any common instances. It must be notices that certain Activities can have several instances within a slice (because of loops and cardinalities of type "many"). We use a slightly different visual representation for this type of Activities for better perception as can be seen in Fig. 2.

When the process is described, is it very important for the doctor to be able to see the run-time instances (both within a slice and over several slices) with their respective attribute values from different points of view. One idea here could be to export all slices over some period of time to Microsoft Excel and then use its features to analyze the data. The main problems here arise from the fact that we can have loops and cardinalities of type "many" allowing several run-time instances appear for a concrete Activity. Developing a non-trivial query for this case may involve serious "Excel programming" not being possible for a doctor. To overcome this problem, we have developed a simple process query language that is based on the process diagram that needs to be analyzed.

4   Process Query Language

The Process Query Language (PQL) has been based on MEDMOD process definition language. The purpose of PQL is to allow a doctor interested in clinical processes querying (filtering) runtime data of hospital's processes described using MEDMOD. In fact, a doctor should be able to ask questions like *"How much did the Dr. Jekyll's patients cost?"* or *"Which patients with Pneumonia had more than two X-rays?"*. This paper describes general ideas behind PQL and does not touch any implementation details. We assume that technical problems, like the import of runtime data from hospitals information system to MEDMOD data structures, have been already solved.

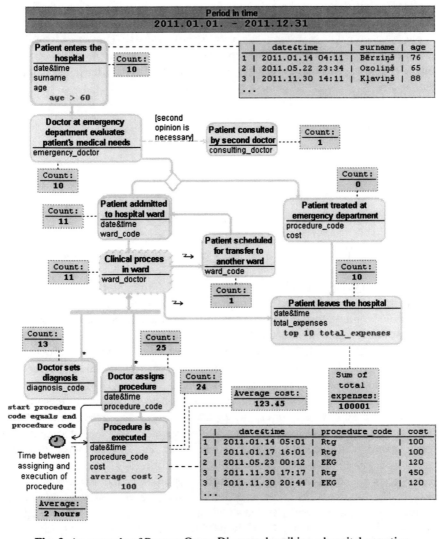

Fig. 3. An example of Process Query Diagram describing a hospital operation

Asking questions begins with choosing (opening) the MEDMOD process diagram, which describes the process under inspection, switching to the **filtering mode** (for example by pressing on a special toolbar button) and setting the time interval the doctor is interested in. As a result, a new diagram – *Process Query Diagram* – is created. It contains the chosen process description in the MEDMOD syntax. In addition, every activity node in the diagram has an indicator (the attached box) showing the number of instances in *the initial dataset* – all slices corresponding to the chosen time interval (see Fig. 3 – an example where the details described in this section can be viewed).

Now the doctor can undertake two types of actions – she can **set filtering conditions** or **retrieve data**. Setting filtering conditions can be initiated by selecting an action node. Typically, the doctor can choose to set a filter on attribute values of the node. The attribute can be selected, for example, by clicking on it. There are several options for filtering. The first filtering option is **the comparison operations** like *equals, greater than, less than, contains, begins with,* etc. The actual list of operations depends on the data type of the attribute. The same principle has been used in spreadsheet applications like *Microsoft Excel* for setting simple filtering conditions on column values. The typical filter input form has been shown in Fig. 4.

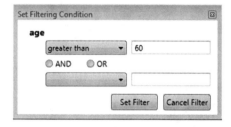

Fig. 4. Filtering condition input form – comparison operation on attribute

First, the comparison operation is selected (e.g., *greater than*). Second, a value is given. If the possible values can be retrieved from the fixed list (e.g., HL7 codes or doctors of the hospital), then the input form offers a list (e.g., via a combo box) the user can choose from. Following the simplicity of spreadsheet applications, only one extra comparison operation is allowed here. User may choose one of the following options – either both conditions are mandatory (logical AND), or at least one of the conditions must be met (logical OR). Thus, most of the typical conditions, including value intervals, can be given using such input form. In the process query diagram a filtering condition appears as a label in the corresponding activity node. Thus, the doctor is always aware of filtering conditions that have been set. Immediately after the filtering condition has been created or updated, it is applied on the dataset. The filtered dataset contains all instances from those slices, which contains instances conforming to the filtering condition. As a consequence, all data displayed in the diagram (e.g., the indicators of number of instances) are updated. The order, in which filtering conditions are set, does not matter – they all are applied every time a new condition has been added or old one deleted.

The second filtering option is **the data partitioning operations** like getting *Top* or *Bottom* instances based on some attribute. Doctor may ask for 10 slices, where *total*

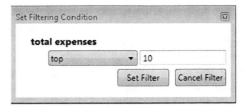

Fig. 5. Filtering condition input form – data partitioning operation

expenses are the largest. She should select the corresponding activity and choose the data partitioning option. The filter input form is shown in Fig. 5.

First, the partitioning operation is selected (e.g., *Top*). Second, a value is given. When the filter is applied, the filtered dataset contains only instances from those slices, which contain instances with ten largest *total expenses* values. It should be noted that partitioning operations are applied on the dataset retrieved by applying the comparison operations. If several data partitioning operations have been set, then the result is the intersection of slices retrieved by partitioning operations.

If there is a possibility that a slice may contain more than one instance of the same type (e.g., if there is a composition with a cardinality "many" or flows heading backwards), then it is possible to set a filtering condition on aggregate functions. The filtering conditions may be applied to the *Sum* or *Average* of attribute values of instances in the slice. The filtered dataset contains all instances from those slices, which contains instances of the filtered type having sum or average value of the given attribute within values specified by the condition. For example, the doctor may ask for those slices, where average *cost* of *"Procedure is executed"* is greater than 100. Another option is to set comparison condition on number of instances within a slice using the *Count* operation. For example, doctor can ask for slices where *"Doctor sets diagnosis"* happened more than once. Setting this condition may be initiated by clicking on the action node itself (not on an attribute).

As stated before, there are two types of querying actions – setting filtering conditions and retrieving data. The former is explained above. Now let us proceed to the latter. Although the MEDMOD diagram elements cannot be modified in PQL, user may supplement the diagram with additional details. The possible options appear in the palette and can be dragged into the diagram, thus retrieving data within the diagram.

The first option is **the time interval**. It can be drawn between actions containing *date&time* attribute. The time interval symbol appears in the diagram. The incoming arrow denotes interval's start activity and the outgoing arrow denotes interval's end activity. The interval has also a name. In fact, we may think of a time interval as of a derived attribute in the master activity of the slice, which is computed as the difference between interval's end activity *date&time* and interval's start activity *date&time*. Note that there may exist multiple interval values because of multiple start and end instances within the slice. To specify more precisely the instances the time interval should be measured between, a conditional expression may be used. For example, if the doctor wants to measure time between *"Doctor sets procedure"* and *"Procedure is executed"* for those instances, whose *procedure code* matches, she should supply the conditional expression stating *"start procedure code equals end procedure code"*. If no conditional expression has been supplied, then the interval between two adjacent (in time) instances of corresponding types is measured. Once the time interval has been defined, it can be used in filtering conditions.

An important feature is the possibility to add intervals between instances allocated in different slices, e.g., between patient's multiple appearances in the hospital. It would require **grouping of slices** to be introduced. Grouping would allow merging slices depending of some attribute values, e.g. patient's *surname*.

The second data retrieving option is the set of **aggregate functions**, which can be evaluated over the filtered dataset in order to obtain a single number as an answer to the question asked. They are: **Count, Sum** and **Average** meaning respectively the number of instances within the dataset, the sum of the given attribute values over all instances in the dataset and the average of the given attribute values over all instances. They can be applied by dragging the selected function from the palette to the corresponding activity node (for *Count*) or to the attribute (for *Sum* and *Average*). The result (one number) appears on the diagram as an indicator box, which displays the computed value. In fact, the number of instances of each activity appears in the diagram by default. However, they can be also removed from the diagram.

The third possible option of retrieving data is a **list of all instances** corresponding to the selected activity. Dragging corresponding palette element to an activity initiates the display of all instances of corresponding type in the filtered (or initial if no filtering conditions are applied) data set. They are displayed as a table, where each row represents an instance and columns represent the attribute values. There is also one special column containing slice's ID the instance belongs to. Since it is possible to display several *instance tables* at once, the presence of an ID in each of them helps to recognize data from the same slice across several tables. If the filtering conditions have been changed, then the content of all tables and indicator boxes is recomputed.

Thus, the basic steps in querying are:

1. Process Query Diagram is created from a MEDMOD diagram – the initial dataset is determined by initial time interval given by the doctor and the indicator boxes denoting the number of instances for each action appear;
2. Doctor may apply two types of filtering conditions – comparison operations on attributes and aggregation functions or data partitioning operations;
3. Doctor may retrieve data into the diagram – aggregate values (Count, Sum, Average), which are one number answers, or instance tables;
4. Changing filtering conditions immediately reflects on displayed data.

The main advantages of the PQL are: 1) the view on data through "glasses" of familiar process, 2) the simple and easy-to-perceive means of setting filtering conditions require no more expertise than using spreadsheat applications (like *MS Excel*), and 3) the dynamic response to each step in construction of the complete query – the doctor sees immediate reaction to every action. It shortens the learning curve greatly and encourages even non-experienced users to try this out.

5   Conclusions and Future Work

To test the practical aspects of using this methodology we presented it to a group of seven doctors working in a hospital. Our primary interest was to assess the "readability" of the designed clinical process model and of the information filtered with its application by the end-users. After a short instruction about the syntax of process description, available filtering mechanisms and visualizing the retrieved information in data indicator boxes next to each of activity nodes, doctors were asked to explain the meaning of the three prepared screenshots representing retrieved data with a use of the query language. All participants of this test demonstrated that they could accurately retrieve the question to be asked by applying the proposed querying techniques in our hospital model. In general, all of the participating doctors rated the presented methodology positively and noted not only the potential for this tool to facilitate management and improve the transparency of clinical processes, but also its potential for research on the impact that certain variables have on the treatment outcomes.

We are about to continue developing this project with two further steps. First, we are planning to develop user-friendly graphical editors for the MEDMOD process modeling and query languages. We have already built prototypes of these editors, which were used in creating the proof of concept, e.g., examples seen in Figures of this paper. Nowadays, it is a rather easy task to develop such domain-specific languages within some of the tool building platforms like GRAF [20] or METAclipse [21].

Our second plan is to develop an effective implementation of the query language. Success of the PQL depends mainly on the efficient implementation of the query execution. This task, is closely related to the pattern matching problem [22] in the field of implementation of model transformation languages (like, MOLA [23], lQuery [24], etc.), which have already been used in the various areas of Model-Driven Engineering.

Acknowledgments. This work has been partially supported by the European Regional Development Fund within the project Nr. 2010/0325/2DP/2.1.1.1.0/10/APIA/VIAA/ 109 and by the Latvian National Research Program Nr. 2 „Development of Innovative Multifunctional Materials, Signal Processing and Information Technologies for Competitive Science Intensive Products" within the project Nr. 5 „New Information Technologies Based on Ontologies and Model Transformations".

References

1. Drucker, P.F.: Managing in the Next Society. Journal of Documentation 59, 209 (2002)
2. Hillestad, R., Bigelow, J., Bower, A., Girosi, F., Meili, R., Scoville, R., Taylor, R.: Can electronic medical record systems transform health care? Potential health benefits, savings, and costs. Health Affairs (Project Hope) 24(5), 1103–1117 (2005)
3. Chaudhry, B., Wang, J., Wu, S., Maglione, M., Mojica, W., Roth, E., Shekelle, P.G.: Improving Patient Care Systematic Review: Impact of Health Information Technology on. Annals of Internal Medicine 144(10), 742–752 (2006)
4. Goldzweig, C.L., Towfigh, A., Maglione, M., Shekelle, P.G.: Costs and benefits of health information technology: new trends from the literature. Health Affairs (Project Hope) 28(2), w282–w93 (2009)
5. Barzdins, J.: Developing health care management skills in times of crisis: A review from Baltic region. International Journal of Healthcare Management 5(3), 129–140 (2012)
6. Burns, L.R., Bradley, E.H., Weiner, B.J., Shortell, S.M.: Shortell and Kaluzny's Health Care Management Organization Design and Behavior. In: Management. Delmar Cengage Learning, Clifton Park (2012)
7. Clark, J., Armit, K.: Leadership competency for doctors: a framework. Leadership in Health Services 23(2), 115–129 (2010)
8. Edwards, N.: Doctors and managers: building a new relationship. Clinical Medicine 5(6), 577–579 (2005), http://www.ncbi.nlm.nih.gov/pubmed/16411354 (retrieved)

9. Sager, A., Socolar, D.: Health Costs Absorb One-Quarter of Economic Growth, 2000-2005. Health (San Francisco), Boston (2005), http://dcc2.bumc.bu.edu/hs/healthcostsabsorbone-quarterofeconomicgrowth2000-05sager-socolar7february2005.pdf

10. Scheuerlein, H., Rauchfuss, F., Dittmar, Y., Molle, R., Lehmann, T., Pienkos, N., Settmacher, U.: New methods for clinical pathways-Business Process Modeling Notation (BPMN) and Tangible Business Process Modeling (t.BPM). Langenbeck's Archives of Surgery / Deutsche Gesellschaft für Chirurgie 397(5), 755–761 (2012)

11. Rojo, M.G., Rolón, E., Calahorra, L., García, F.O., Sánchez, R.P., Ruiz, F., Espartero, R.M.: Implementation of the Business Process Modelling Notation (BPMN) in the modelling of anatomic pathology processes. Diagnostic Pathology 3(suppl. 1), S22 (2008)

12. Müller, R., Rogge-Solti, A.: BPMN for Healthcare Processes. In: Proceedings of the 3rd Central-European Workshop on Services and Their Composition, ZEUS 2011, Karlsruhe, Germany, February 21–22, pp. 65–72. CEUR-WS.org, Karlsruhe (2011), http://ceur-ws.org/Vol-705/paper9.pdf (retrieved)

13. Agt, H., Kutsche, R.-D., Wegeler, T.: Guidance for domain specific modeling in small and medium enterprises. In: Proceedings of the Compilation of the Co-Located Workshops on SPLASH 2011 Workshops, vol. 63 (2011)

14. Beeri, C., Eyal, A., Kamenkovich, S., Milo, T.: Querying business processes. In: VLDB, pp. 343–354 (2006)

15. Awad, A.: BPMN-Q: A Language to Query Business Processes. In: EMISA, pp. 115–128 (2007)

16. Beeri, C., Eyal, A., Milo, T., Pilberg, A.: Monitoring business processes with queries. In: Proceedings of VLDB 2007, Vienna, Austria, pp. 603–614 (2007)

17. Beheshti, S., Benatallah, B., Motahari-Nezhad, H.R., Sakr, S.: A Query Language for Analyzing Business Processes Execution. In: Rinderle-Ma, S., Toumani, F., Wolf, K. (eds.) BPM 2011. LNCS, vol. 6896, pp. 281–297. Springer, Heidelberg (2011)

18. Bārzdiņš, J., Bārzdiņš, G., Čerāns, K., Liepiņš, R., Sproģis, A.: UML Style Graphical Notation and Editor for OWL 2. In: Forbrig, P., Günther, H. (eds.) BIR 2010. LNBIP, vol. 64, pp. 102–114. Springer, Heidelberg (2010)

19. Health Seven Level International, http://www.hl7.org

20. Sproģis, A., Liepiņš, R., Bārzdiņš, J., Čerāns, K., Kozlovičs, S., Lāce, L., Rencis, E., Zariņš, A.: GRAF: a Graphical Tool Building Framework. In: Proceedings of the Tools and Consultancy Track. European Conference on Model-Driven Architecture Foundations and Applications, Paris, France, pp. 18–21 (2010)

21. Kalnins, A., Vilitis, O., Celms, E., Kalnina, E., Sostaks, A., Barzdins, J.: Building Tools by Model Transformations in Eclipse. In: Proceedings of DSM 2007 Workshop of OOPSLA 2007, pp. 194–207. Jyvaskyla University Printing House, Montreal (2007)

22. Sostaks, A.: Pattern Matching in MOLA. In: Barzdins, J., Kirikova, M. (eds.) Proceedings of the 9th International Baltic Conference on Databases and Information Systems, Riga, Latvia, July 5-7, pp. 309–324. University of Latvia Press, Riga (2010)

23. Kalnins, A., Barzdins, J., Celms, E.: Model Transformation Language MOLA. In: Aßmann, U., Akşit, M., Rensink, A. (eds.) MDAFA 2003. LNCS, vol. 3599, pp. 62–76. Springer, Heidelberg (2005)

24. Liepiņš, R.: Library for Model Querying – lQuery. In: Proceedings of 2012 Workshop on OCL and Textual Modelling (2012)

Towards Dynamic Non-obtrusive Health Monitoring Based on SOA and Cloud

Mohamed Adel Serhani, Abdelghani Benharref, and Elarbi Badidi

College of Information Technology, UAE University
UAE, Al-Ain
{serhanim,abdel,ebadidi}@uaeu.ac.ae

Abstract. Despite the fact that new technologies and life style are continuously improving, many diseases have unfortunately extensively increased in today's population. State-of-the-art studies are showing an exponential increase of these diseases, which present a heavy burden on governmental and private healthcare systems. Many industrial and academic works are trying to alleviate this burden using varying clinical solutions. For example, e-health monitoring and prevention have revealed to be among promising solutions. In fact, well-implemented monitoring and prevention schemes have resulted in a decent reduction of diseases risk and or have reduced their effects. In the same line, this paper proposes a Service-Oriented Architecture-based platform for health disease tracking and prevention with a focus on disease monitoring. A monitoring scheme based on Service-Oriented Architecture and Cloud technology has been designed and developed to proactively detect any risk of diseases prior to its development. Therefore a preventive plan is dynamically generated and customized according to the patient's health profile and context while considering many impelling parameters. A mobile application has been developed to evaluate our monitoring scheme and preliminary data have been collected and analyzed.

Keywords: E-health, Diseases, Monitoring, Prevention, SOA, Cloud.

1   Introduction

The last decade has witnessed a high and increasing ratio of deaths caused by chronic and cardiovascular disease (CVD) in all countries over the world [1]. For example, and according to the World Health Organization (WHO), the UAE has the second highest rate of diabetes in the World. Moreover and according to the statistics from the United Arab Emirates (UAE) Ministry of health published in 2010, over 25% of deaths in the country are caused by cardiovascular disease [2]. Diabetes, high blood cholesterol and other lipids, physical inactivity, smoking, overweight and obesity are some factors that can cause these diseases.

To address the rising incidence of diseases and their associated complications, a continuous monitoring-driven prevention approach can contribute to explaining the main reasons of HDs and then reducing the risk of their occurrences. Also, continuous monitoring of subjects is a very important element in detecting diseases' symptoms as soon as they occur; therefore, it can mitigate the impact and the consequences that

G. Huang et al. (Eds.): HIS 2013, LNCS 7798, pp. 125–136, 2013.
© Springer-Verlag Berlin Heidelberg 2013

these diseases may cause. Tracking is also an essential mechanism in locating the subject while he/she is away of usual known points.

Western healthcare industry is undergoing fundamental changes. It is shifting from hospital-centric services to a more ubiquitous and ambulatory system (with homecare, day care clinics, remote healthcare) and the treatment of diseases that actively involves the patient himself/herself. The emergence of Web-based e-Health services allows patients and professionals to have easy access to important information anytime anywhere while optimizing healthcare cost. The UAE Ministry of Health has recently announced the implementation of Wareed health system that will connect hospitals and clinics in the UAE to improve related services and reduce costs [3].

Application of the above mechanisms requires full integration of different technologies, systems, and communication infrastructures. Continuous on-the-fly monitoring requires appropriate devices, such as mobile sensors, in addition to pervasive computing; communication with hospitals and physicians requires wireless/mobile-networking technologies. These technologies provide the foundation to support efficient and continuous HDs prevention, tracking, and monitoring.

In this work, we propose a non-invasive mobile health monitoring architecture, for monitoring of diseases. The system relies on the concept of Service Oriented Architecture (SOA) and Cloud computing. SOA has been proven to be an adequate solution for integrating heterogeneous systems and technologies, allowing application-to-application communication over the Internet, reducing cost of integration, and exposing data and services to different stakeholders. The system will allow continuous data gathering, automatic monitoring, and taking proactive measures to identify potential risky situations and prevent the subject from severe heath consequences. Besides, due to the use of standard and open technology, the system can be easily integrated with other healthcare systems, which allows high interoperability and dynamic integration between heterogeneous systems.

The monitoring approach, to detect HDs as soon as they occur, involves the usage of sensing technologies along with communication infrastructure to sense and collect important physiopathology measures of a patient. It addresses topics such as measuring and collecting sensory information, purging and enhancing this sensory information, and making intelligent and health-critical decisions. Major discrepancies in measured data triggers immediate analysis and assessment of the subject's data by physicians to find out whether the subject has or not a chronic disease.

The preventive approach to reduce the incidence of HDs and control the spread of these diseases, involves addressing the causing factors mainly obesity, hypertension, smoking, diabetes, overweight, dyslipidemia, physical inactivity and unhealthy diet and life style. Collection, analysis, and mining of data are key elements in addressing the problem. It consists of extracting patterns to build prediction models that greatly help taking rational decisions to prevent and decrease the risk factors.

In this paper, we propose an architecture for diseases monitoring. The system is based on the SOA paradigm [4] , advanced information and communication technology (ICT) and Cloud infrastructure, platform and applications. Adopting SOA will allow developing a pool of e-health services that will ease the integration with other healthcare systems. The proposed architecture will also allow the integration and the usage of these services from mobile devices.

The remaining sections of this paper are organized as follows: next section discusses the state of the art in e-health and m-health monitoring. Section 3 discusses our

Smart and Non-Invasive Health Monitoring architecture based on SOA and Cloud and SOA. Section 4 presents our smart and non-invasive health monitoring based on cloud and SOA. Section 5 concludes the paper and highlights some of future works.

2   Related Work

Nowadays, several e-health centers emerged and many initiatives are launched in USA, UK, EU, AU as well as in many developing countries in order to improve healthcare services and optimize medical resources. At the core of all this initiatives is the investment in modern Internet and Communication Technology (ICT) infrastructures to connect hospitals, clinics and healthcare organizations so that the exchange of medical data can be possible.

The increasing demand for e-Health services has led to many research efforts [5]. Different architectures have been proposed for e-Health or other healthcare related system purposes. Some of them are used in special areas, such as trauma [6], and cardiology [7]. Some are used with special purposes, such as emergency and patient monitoring [8, 9]. With the advantages of wireless and mobile technologies, there are also many wireless-based e-Health systems presented in [10]. The common feature of these systems is that they only provide limited or special services to users.

Several research works and initiatives have investigated the issue and challenges of building e-health solutions. These solutions differ mainly in the way they handle integration issues due to heterogeneity of different systems and type of middleware, framework or architecture used to build an integrated e-health system. The authors in [11] proposed a distributed framework of Web-based telemedicine system, which addresses two types of servers: 1) Web servers and 2) data servers. This framework is based on CORBA technology and a database. The system is not flexible to allow any integration of non-CORBA systems; therefore, it requires an intermediary middleware to handle the heterogeneity between different heath systems as well as a huge development effort to adapt the system to the integrated system requirements.

In [12], authors proposed a multi-layer SOA-based e-Health services architecture that consists of six main components responsible for defining interactions among different layers. The system is generic; it states an architectural design without detailing the implementation and its challenges. In addition, the proposed system has not been implemented. Kart, F. et al. [13] described a distributed e-healthcare system that uses SOA as a means of designing, implementing, and managing healthcare services. The users of the system are physicians, nurses, pharmacists, and other professionals, as well as patients. The system includes a clinic module, a pharmacy module, and patient's interfaces, which are implemented as Web services. Various devices can interact with these modules, including desktop and server computers, PDAs and smart phones, and even electronic medical devices, such as blood pressure monitors.

The authors in [14] described the design, the implementation, and the deployment of a multi-tier Inpatient Healthcare Information System based on SOA and on the HL7 massage exchange standard at the National Taiwan University Hospital (NTUH). The services-tier includes Computerized-Physician Order Entry (CPOE), Billing, Pharmacy, and Diet. This work also investigates how healthcare organizations, using SOA, can leverage their shared services to automate multiple business processes and reinforce overall interoperability. The authors in [15] designed and developed a SOA-based platform for home-care delivery to patient with diseases. Always on monitoring, it presents applications and requirements of pervasive healthcare, wireless networking solutions and several important research problems. These healthcare applications include pervasive health monitoring, intelligent emergency management system, pervasive health- care data access, and ubiquitous mobile telemedicine.

To promote interoperability among healthcare organizations that are seeking to develop SOA-based architectures, a joint collaboration effort among standards groups, specifically Health Level Seven (HL7) and the Object Management Group (OMG), was formed under the name: Healthcare Services Specification Project (HSSP). The intent is to produce SOA health standards standard services that define services' responsibilities, behavior, and interfaces so that ubiquity can be achieved across implementations and vendor products [16].

Our solution is aligned with the above initiatives and addresses mainly diseases monitoring. It also addresses some difficult issues in the design of an e-health system and protection of medical data. Our solution relies on SOA and the Cloud to integrate different systems, data, and make it available for the sake of diseases monitoring anywhere and whenever needed. The potential of using SOA in our solution is that it facilitates interoperation among various systems that typically do not speak the same language. Using SOA as a common standard enables reducing the complexity in integrating heterogeneous systems. New services can be developed to satisfy the needs of integration, and existing system capabilities can also be organized into services. Redundant processing, which is normally developed and used by several units, can be structured and represented as a separate service or set of services. Each service becomes, then, available to all stakeholders through a standard interface. Furthermore, services can be orchestrated in such a way to be aligned with users' workflows.

Beside the challenges in the development and implementation of the above model, several research issues are widely open, in particular the collection of appropriate data using appropriate sensors and at appropriate intervals, integration of different technologies and systems to solve the interoperability are among the difficult challenges. Being able to solve these issues and build such system will be a breakthrough in research and development of complete safety-critical m-Healthcare systems.

3 Health Monitoring Architecture

3.1 Architecture Description

In this section, we describe our architecture that is founded on cloud infrastructure and the open SOA. Figure 1 depicts the proposed architecture based on our initial investigations on the issue of continuous and accurate monitoring of patient's health data. The particularity of our solution is its full integration of highly promising sensing and monitoring technologies, cloud infrastructure, SOA to enable smart monitoring of patient's vital health parameters. The architecture categorizes varying contributors within the solution along with their contribution and participation in the monitoring, evaluation, and decision processes. Subsequent section describes each module and actor in addition to their main roles and duties within the architecture.

Non-Invasive Sensing Layer: this layer includes any device (e.g. sensor, mobile) that can be used to sense one or more health parameters such as blood pressure, blood sugar, body temperature, oxygen saturation, etc. It also includes gateways that serve in collecting data from sensors, perform some processing (e.g. filtering) and store these data on the Cloud. Many sensing technologies are available nowadays and used to retrieve vital information, and then relay these to the nearest device capable (e.g. gateway) to process these health data. These intermediate relays generally provide an interface to access and retrieve these data.

Cloud Application, Platform, and Infrastructure Layer: this layer serves as the underlying infrastructure and platform that hosts e-health data and applications. This includes connectivity management, device management, e-health data processing, and other cloud based services that support monitoring activities and related processes.

Access and Integration Layer: this layer is responsible for integrating other healthcare systems with the monitoring architecture as well as providing the necessary interfaces to allow an easy access and manipulation of heath data. The main component of this layer is the Enterprise Service Bus (ESB) that allows interoperation and exchange of data among different sub-systems. Other components of this layer include Web services interfaces that can be invoked by service consumers in order to explore and process monitoring data. This layer also manages process integration using workflows and service composition.

Presentation Layer: this layer consists a set of applications and services that might be used for collection and processing of health data. Type of services might include report generation, pattern mining and recognition tools, and data visualization.

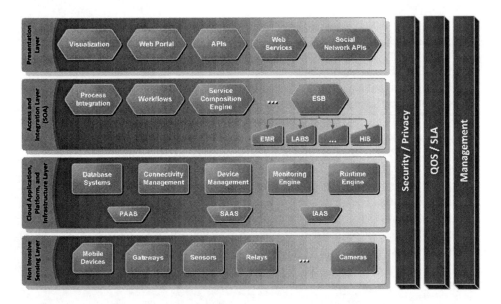

Fig. 1. SOA & Cloud enabled Architecture for Smart Non-Invasive Health Monitoring

The other three towers of the architecture include Security mainly privacy, QoS-SLA, and management towers. These areas are applied in all the four architecture layers. Security and privacy of the data, services and applications is of prime importance as it deals with private patient's health data. QoS-SLA includes establishing and guarantying service level agreement with all roles and partners in the architectures in terms of QoS enforcement and violation. Management is of prime importance for all the architecture components and layers; this includes management processes that span from sensing devices management to service and application management.

3.2   Monitoring Lifecycle

In this section, we will illustrate the main processes of smart health monitoring. The monitoring lifecycle involves a set of roles and components that include the patient, physicians (healthcare professionals), monitoring platform (mobile, Sensors), and Engine system.

Patient: can be: (1) any person suffering from a health disease and can use the monitoring system to benefit from continuous monitoring of his/her disease, and to get fast assistance and intervention by physicians whenever a need raises, (2) any person that do not suffer from any diseases but who is interested in using the system to monitor periodically his/her health condition.

Sensing platform: includes any sensing devices, equipment, and gateways that are used to sense the patient's health parameters and send real time data to a backend server. The latter includes a smart engine that processes, mines and filters these data to detect and report if any discrepancy of patient's health data is triggered. Therefore, healthcare professionals are informed with the report monitoring status to take actions, if needed.

Expert system (Engine): is in charge of processing patient's collected data and execute some data mining operations. The ES receives and analyses notification messages that might be sent by the gateway if an expected heath condition of a subject has occurred. The ES also handles SOS notification messages to the healthcare professionals when an emergent situation is detected by transmitting the details of the situation and the monitored heath data of the patients. Also, the ES exhibits set of services that are made available to physicians and to patient's assistance team. These services expose data such as laboratory tests, demographic, anthropometric, and biological data. This information serves as decision support for physicians to take the appropriate actions.

Healthcare professionals: includes various physicians and healthcare professionals that might be involved in the examination of monitored subject. After receiving SOS messages from the ES, it triggers necessary actions and notifies physicians of the current conditions of the subject under observation and provides them with the necessary data. Depending on the detected health problem (disease) expert physicians on the disease domain are called to respond to the case and follow up with the patient's heath situation. Physicians are supported by the monitored data they received in addition to other laboratory test, scanners, historical data, etc.

4   Smart Health Monitoring

In addition to the above roles, our model exhibits two key features that concern (1) visualization of assistance data, statistics and report generation, data mining and patterns detection, (2) prevention and action plan generation to respond to the occurrence and treatment of diseases. A prevention plan may recommend, for instance, practicing regular sport exercises, following a diet plan, changing the food habits and the lifestyle, etc. An action plan, however, consists of a series of actions that might include medications, chemotherapy, reeducation, etc.

Figure 2 shows how blood pressure parameter is monitored and data is sent to a mobile device, afterwards it is transmitted to a cloud data center. Once the data is available on the Cloud surveillance center in a hospital for instance, it can access these data and notify the appropriate health professionals if an intervention is required

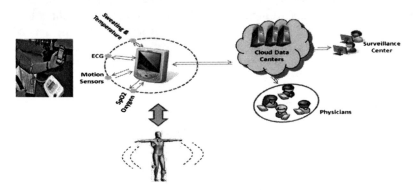

Fig. 2. Monitoring process illustration

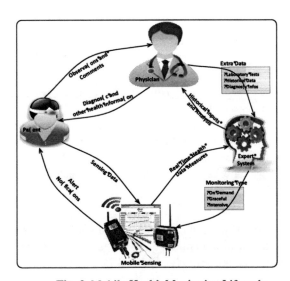

Fig. 3. Mobile Health Monitoring Lifecycle

in response to any inconsistency of monitored vital parameters. Also, physicians can access the historical monitoring data of monitored patient stored on the cloud for further investigation purpose.

Figure 3 describes the mobile health monitoring process and the main component involved in the accomplishment of this process. It summarizes the interactions among system actors.

5   Implementation

To illustrate the proposed monitoring scheme, we present the development of a mobile health monitoring application that incorporates three monitoring scenarios that we have proposed for the purpose of varying patients' contexts. This application adopts a self-adaptive non-invasive health-monitoring scheme in which the observed patient is not aware of monitoring activities that are taking place. In the following, we describe the environment setup, the monitoring scenarios, the mobile application we have developed for the sake of monitoring, and the experimentation results we have obtained along with their interpretations.

5.1   Environment Setup and Main Scenarios

We have conducted a series of experiments to monitor and collect a couple of physiopathology health data. This includes the following parameters: temperature, blood pressure, ECG, heart rate, and blood sugar. To collect these measurements, we setup a wireless sensor network using a set of sensors. These sensors sense real time data and transmit it to a mobile device. Then, the collected data is stored and processed on a Cloud data center. The Cloud also offers some data processing services including filtering, mining, fusion, and visualization.

Setup

To execute the above experiments, we have settled the following environment:

- Samsung mobile Galaxy Note running Android 4 to retrieve sensed data.
- Sensing devices including communication protocol and APIs.
- A Non-Invasive Mobile application developed to parse the sensed data stored on the cloud generates, visualizes, and interprets the monitoring results. Then sends alerts and summary information.
- Cloud data center in which sensed data is stored and managed.

Scenarios

We have developed monitoring scenarios that capture different patient's situations form normal to moderate health condition. These scenarios include monitoring two vital health parameters that are Blood Sugar (BS) and the ECG (Electro Cardio

Gram). The first scenario considers two reading of BS: the first reading is taken while fasting (BSF) and the second reading is relatively around 2 hours after meal (BSM). The second scenario timestamps each ECG reading signals for a monitoring period, which may span over 1 minute.

Sample Data Set

Our experiments used the monitoring dataset (Table 1) taken in different period of the day and in different subject's context (moving, eating, and working).

Table 1. Sample Dataset

	Time(Sec)	Sig 0 (mV)		BSF	BSM
ECG	0.000	0.165	Blood Sugar	117	198
	0.008	0.155		123	220
	0.016	0.195		123	135
	0.023	0.205		129	171
	0.031	0.185		120	121

MM-Health Application

The mobile monitoring application (MM-health) we have developed fetches the sensed data stored on the cloud and offers a set of interesting features. Figure 4 describes a snapshot of the application and its main components. These include monitoring a couple of key vital signs such as Blood Pressure, ECG, Heart Rate, Cholesterol, Blood Sugar, Temperature, and Body Movement. For each of the monitored health

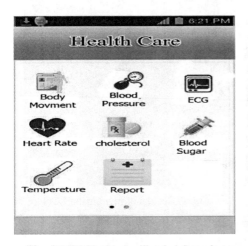

Fig. 4. MM-Health Application Snapshot

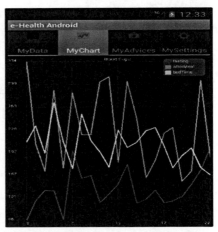

Fig. 5. Results of Monitoring Blood Sugar

parameter a couple of services are offered, and include data visualization (MyData), graph generation (MyChart), automatic advices generation (MyAdvices), and a parameterized monitoring setting (MySettings). In addition, the application provides a reporting service that allows the user to generate different reports and statistics.

We also developed an engine that processes the collected data on the cloud and generates some important information that helped in making prevention plan and/or other clinical decision. These processed data is visualized on the MM-heath application. The main features of the application include: collecting sensed data, provide a health status, health data visualization, continuous awareness messages generation, alert and reminder, share health information, prevention plan, and other important information.

Results and Interpretation

Figure 5 shows a graph resulting from monitoring the blood sugar in three situations: fasting, after meals, and at bedtime. The mobile application also incorporate a summary of test results in addition to some advices generated for the patient based of the Blood Sugar observed levels as illustrated in Figure 6. These advices include for instance practicing sport, and consulting urgently a physician. Figure 7 shows the results of monitoring the ECG; the user can scroll horizontally the screen to see the rest of ECG signal. No interpretation of results is made at this stage, as it is hard to automate the interpretation of the ECG signal. It is subject to other ECG parameters, and needs a physician intervention.

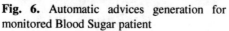
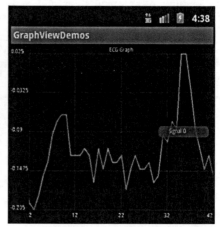

Fig. 6. Automatic advices generation for monitored Blood Sugar patient

Fig. 7. Results of monitoring ECG

From the scenarios we have executed, we have drawn the following conclusions:

- Mobile monitoring is a very challenging process as it is subject to many restrictions; this includes mobile device limited processing capacity, unpredictability of network connection, and power drainage of both sensor and mobile device.
- Automatic diagnosis and interpretation of health data cannot be fully automatic as it consists of different parameters such as analyzing patient's historical health data, his/her laboratory test, etc. therefore needs intervention of health professionals.

6 Conclusion

Mobile health monitoring has been recognized as an important action to monitor vital parameters of a subject under observation. This is explained by the mobility of patients and the nature of their work conditions. Many solutions were implemented to mitigate the impact of main diseases via monitoring proactively and intrusively patient's health parameters in order to detect and prevent the consequences these diseases might cause.

We proposed a solution based on a proactive monitoring and scheme supported by an open, flexible, high scalable, and interoperable SOA and Cloud based solution. This architecture allows monitoring of vital parameters of a patient while moving, generates important information and advices visualized on a mobile device and stores on a cloud data center for better processing, maintenance, and privacy guarantee.

As future work, we are planning to conduct extensive monitoring scenarios on large data set and do some data mining of these data which will enable a better understanding of the nature of diseases and evaluate the ability of our solution in guarantying performance, security, and wide range of services.

Acknowledgment. This work has been sponsored by UAE University under the NRF grant #21T020. We would like also to thank the development team who are working on the development of the remaining modules of the architecture.

References

1. Serhani, M.A., Badidi, M.E., Benharref, A., Dssouli, R., Sahraoui, H.: Integration of Management of Quality of Web Services in Service Oriented Architectures. In: Sugumaran, V. (ed.) Intelligent Information Technologies and Applications. Advances in Intelligent Information Technologies (AIIT) Book Series, pp. 190–220 (2007)
2. Benharref, A., Glitho, R., Dssouli, R.: Mobile agents for testing web services in next generation networks. In: Magedanz, T., Karmouch, A., Pierre, S., Venieris, I.S. (eds.) MATA 2005. LNCS, vol. 3744, pp. 182–191. Springer, Heidelberg (2005)
3. Benharref, A., Dssouli, R., Glitho, R., Serhani, M.A.: Towards the testing of composed web services in 3rd generation networks. In: Uyar, M.Ü., Duale, A.Y., Fecko, M.A. (eds.) TestCom 2006. LNCS, vol. 3964, pp. 118–133. Springer, Heidelberg (2006)
4. Papazoglou, M.P., Traverso, P., Dustdar, S., Leymann, F.: Service-oriented computing: State of the art and research challenges. Computer 40, 38–45 (2007)

5. Kaur, G., Gupta, N.: E-health: A new perspective on global health. Journal of Evolution and Technology 15, 23–35 (2006)
6. Chu, Y., Ganz, A.: A mobile teletrauma system using 3G networks. IEEE Transactions on Information Technology in Biomedicine 8, 456–462 (2004)
7. Fayn, J., Ghedira, C., Telisson, D., Atoui, H., Placide, J., Simon-Chautemps, L., et al.: Towards new integrated information and communication infrastructures in e-health. Examples from cardiology. In: Proceedings of Computers in Cardiology, pp. 113–116. IEEE (2003)
8. Jovanov, E.: Wireless technology and system integration in body area networks for m-health applications. In: 27th Annual International Conference of the Engineering in Medicine and Biology Society, IEEE-EMBS, pp. 7158–7160. IEEE (2006)
9. Istepanian, R.S.H., Jovanov, E., Zhang, Y.: Guest editorial introduction to the special section on m-health: Beyond seamless mobility and global wireless health-care connectivity. IEEE Transactions on Information Technology in Biomedicine 8, 405–414 (2004)
10. Xiang, Y., Gu, Q., Li, Z.: A distributed framework of Web-based telemedicine system. In: Proceedings of the 16th IEEE Symposium Computer-Based Medical Systems, pp. 108–113. IEEE (2003)
11. Omar, W.M., Taleb-Bendiab, A.: E-health support services based on service-oriented architecture. IT Professional 8, 35–41 (2006)
12. Hsieh, S., Hsieh, S., Weng, Y., Yang, T., Lai, F., Cheng, P., et al.: Middleware based inpatient healthcare information system. In: Proceedings of the 7th IEEE International Conference on Bioinformatics and Bioengineering, BIBE, pp. 1230–1234. IEEE (2007)
13. Kart, F., Moser, L.E., Melliar-Smith, P.M.: Building a distributed e-healthcare system using SOA. IT Professional 10, 24–30 (2008)
14. Juneja, G., Dournaee, B., Natoli, J., Birkel, S.: SOA in healthcare (Part II). SOA Magazine (2009)
15. Yang, C.L., Chang, Y.K., Chu, C.P.: Modeling Services to Construct Service-Oriented Healthcare Architecture for Digital Home-Care Business. In: International Conference on Software Engineering & Knowledge Engineering, pp. 351–356 (2008)
16. HL7 and OMG, The practical Guide for SOA in Health Care: A real World Approach to Planning, designing, and deploying SOA (2008),
    http://hssp.wikispaces.com/PracticalGuide

Case-Centred Multidimensional Scaling
for Classification Visualisation in Medical Diagnosis

Frank Klawonn[1,2], Werner Lechner[3], and Lorenz Grigull[3,4]

[1] Department of Computer Science
Ostfalia University of Applied Sciences
Salzdahlumer Str. 46/48, D-38302 Wolfenbuettel, Germany
[2] Bioinformatics & Statistics
Helmholtz Centre for Infection Research
Inhoffenstr. 7, D-38124 Braunschweig, Germany
[3] Improved Medical Diagnostics Pte Ltd
190 Middle Road, #19-05 Fortune Centre, Singapore, 188979
[4] Department of Pediatric Haematology and Oncology
Medical University
Carl-Neubergstr. 1, Hannover, Germany

Abstract. Computer-based decision support can assist a medical doctor to find
the right diagnosis. The knowledge and experience of the medical doctor is enhanced by a much larger data set of patients than the doctor will ever see in her
or his life. The decision support system can derive possible diagnoses for a new
patient based on a suitable classifier built on the patients in the patient database.
However, since such a system cannot replace a medical doctor and should only
support her or him, it should also provide information about the certainty of its
recommendation. In this paper, we propose to visualise how close or similar the
new patient is to others in the database by a modified multidimensional scaling
technique that focuses on the correct positioning of the new patient in the visualisation. In this way, the medical doctor can easily see whether the diagnosis
recommended by the system is reliable when all patients close to the new patient
have the same diagnosis or whether it is quite uncertain when the new patient is
surrounded by patients with different diagnoses.

1 Introduction

To arrive at a medical diagnosis is a complex process. Doctors usually include clinical and laboratory findings to generate a hypothesis and then systematically rule out
differential-diagnoses by experience and/or logical thinking. Unfortunately, this process is prone to mistakes. As a consequence, tired doctors or doctors in the emergency
department or unexperienced doctors are at risk to pose wrong diagnoses.

We therefore aimed at developing a tool to support the medical diagnostic decision
process in the paediatric emergency department. Data sets of about 700 patients with 18
different medical diagnoses frequently encountered at a tertiary childrens hospital were
included. Using an ensemble of three different classifiers, a new patient data record
could then be allocated to the correct diagnosis with good reliability thus resulting in a
good diagnostic support for doctors.

G. Huang et al. (Eds.): HIS 2013, LNCS 7798, pp. 137–148, 2013.

The decision support system for the medical doctor can be incorporated at any phase of the examination of the patient. In a very early state, when only a few attributes or measurements of the patient are available, the decision support system might not be able to propose a diagnosis or only one with great uncertainty. With more and more measurements taken from the patient, the decision support system will be able to provide one or two possible diagnoses with high certainty.

It is important for the medical doctor to obtain information about the status of the decision support system, how certain it is about the proposed diagnosis. The system proposed in [1] which we use in this paper as a case study, will provide an A- and a B-diagnosis. The A-diagnosis is the one considered to be most probable, the B-diagnosis is the one with the second highest probability. The system will not provide any diagnosis if the A-diagnosis has too little certainty. Nevertheless, it is crucial to know whether the system is relatively sure about the proposed A-diagnosis or whether the B-diagnosis is almost as likely as the A-diagnosis.

This could be indicated by probabilities or scored. But a visualisation can include more information than a single number. Therefore, we have developed a visualisation technique that shows how close the new patient is to those ones in the database for which the diagnosis is already known.

Our method is based on principles of multidimensional scaling, a visualisation technique briefly reviewed in Section 2. However, the focus of our visualisation should be put on the new patient, not on data visualisation in general. Therefore, we connect multidimensional scaling as a general purpose data visualisation technique with classifiers and classifier ensembles in Section 3. Since the new patient to be classified should be in the centre of the visualisation, we propose a modified version of multidimensional scaling in Section 4 which we call case-centred multidimensional scaling. The algorithm to compute the visualisation for case-centred multidimensional scaling is explained in Section 5. Section 6 illustrates how our visualisation technique works for artificial and real data. The final conclusions summarise our results and emphasize that – although we have focused on medical diagnosis – our visualisation technique can also be applied in other areas.

2   Multidimensional Scaling and Distance-Based Visualisation

Multidimensional scaling (MDS) is a dimension reduction technique which is mainly used for visualisation of high-dimensional data. MDS assumes that a p-dimensional data set $X = \{x_1, x_2, ..., x_n\} \subseteq \mathbb{R}^p$ is given. By $d(v, w)$ we denote the Euclidean distance $\|v - w\|$ between points $v \in \mathbb{R}^p$ and $w \in \mathbb{R}^p$. In MDS each of the high-dimensional data points x_i has to be mapped to a low-dimensional representative y_i. The projection of X is denoted as $Y = \{y_1, y_2, ..., y_n\} \subseteq \mathbb{R}^q$ where $1 \leq q < p$ (typically $q \in \{2, 3\}$). A perfect distance-preserving projection of X to Y would keep the distances $d_{ij}^x = d(x_i, x_j)$ of the high-dimensional space identical to the distances $d_{ij}^y = d(y_i, y_j)$ of the projected data objects, that is, $d_{ij}^x = d_{ij}^y$ holds. A perfect projection is, however, impossible except for a few trivial cases. Therefore, MDS seeks to minimise the error introduced by the projection ($|d_{ij}^x - d_{ij}^y|$ for all i, j). Common objective functions for MDS are [2]:

$$E_1 = \frac{1}{\sum_{i=1}^{n} \sum_{j=i+1}^{n} \left(d_{ij}^x\right)^2} \sum_{i=1}^{n} \sum_{j=i+1}^{n} \left(d_{ij}^y - d_{ij}^x\right)^2, \tag{1}$$

$$E_2 = \sum_{i=1}^{n} \sum_{j=i+1}^{n} \left(\frac{d_{ij}^y - d_{ij}^x}{d_{ij}^x}\right)^2, \tag{2}$$

$$E_3 = \frac{1}{\sum_{i=1}^{n} \sum_{j=i+1}^{n} d_{ij}^x} \sum_{i=1}^{n} \sum_{j=i+1}^{n} \frac{\left(d_{ij}^y - d_{ij}^x\right)^2}{d_{ij}^x}. \tag{3}$$

E_1 is based on the absolute errors for the distances, E_3 on the relative errors and E_2 is a compromise between the absolute and the relative error. E_2 is also called stress and MDS based on E_2 is called Sammon mapping. The minimisation of any of the objective functions for MDS poses a non-linear optimisation problem, so that the selected objective function is usually minimised by a numerical optimisation technique.

It should be noted that MDS only needs the distances d_{ij}^x between the data points, not the coordinates of the points themselves. MDS does not have to be based on the Euclidean distance. Any other distance measure between the data points can also be used to define the values d_{ij}^x. This fact will also be exploited in the following section when we explain the connection MDS and the visualisation of a classification result.

3   Visualisation of Classification Decisions

In this paper, a modified version of MDS is introduced, not for the primary purpose of visualising the patients in a database for whom the diagnosis is already known, but for illustrating how similar a new patient is to other patients in the database.

Figure 1 uses an artificial data set to illustrate how such a visualisation could look. The red cone represents the new patient. Patients in the database with diagnosis A are represented by green spheres, whereas patients in the database with diagnosis B are marked by blue cubes. In this case, the new patient fits very well to the patients with diagnosis A and is not very close to the patients with diagnosis B.

But how would one generate such a visualisation based on MDS? MDS requires the pairwise distances between the patients in the database as well as distances of the new patient to the patients in the database. One could simply use the Euclidean distance as a distance measure. This would require that all attributes are numerical, so that categorical attributes would have to be converted into numerical attributes. Furthermore, in order to be independent of the influence of the measurement unit, each attribute should be normalised in a suitable way, for instance by scaling its range to the unit interval. Conversion of categorical attributes to numerical ones and normalisation is not the topic of this paper and we refer to a more detailed discussion on these topics to [3].

Missing values must also be taken into account. Missing values in patient data are quite normal, since not all possible measurements will be taken from each patient. The Euclidean distance between two patients could be computed based only on those attributes without missing values. But this would introduce a bias to smaller distances for

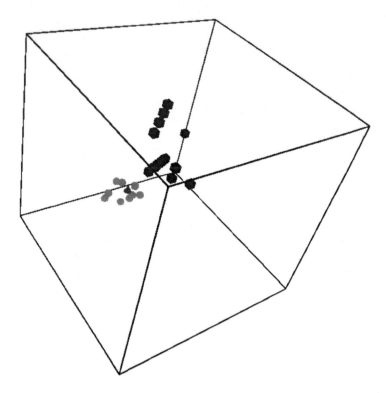

Fig. 1. Visualisation for an artificial data set where the classes or diagnoses are well separated and the new patient is in the centre of patients from one diagnosis. The cube is drawn to provide a better 3D perception.

patients with a larger number of missing values. To avoid this bias, this distance could be divided by the number of attributes which where available for the distance calculation. Instead of the Euclidean distance, suitable kernel-based distances [4] could also be considered. However, a visualisation based on the Euclidean or a kernel-based distance would essentially illustrate how well a nearest neighbour classifier would perform.

But the system for diagnostic support might not be based on a nearest neighbour classifier, but on other classification techniques. In our case [1], none of the single classifiers we tried showed a fully satisfactory performance. Therefore, we rely on a classifier ensemble [5] of three classifiers: a support vector machine, a neural network and a fuzzy rule-based system.

The visualisation of the performance of a classifier ensemble could be based on the distances between the outputs of the different classifiers instead of the Euclidean distances based on the attributes of the patients. This idea is illustrated by Table 3. For the new patient and also for each patient in the database, we have k different possible diagnoses and each of the c classifiers of the classifier ensemble provides a score for each of the diagnoses. This means, each patient is represented by a $(k \cdot c)$-dimensional vector. The Euclidean distance between these vectors could be used as basis for the MDS visualisation. In our specific application, we have 18 diagnoses and 3 classifiers, so that the

Table 1. Structure of the data table for the distance calculation based on the results of the single classifiers

Patient	Classifier 1			...	Classifier c		
	Diagnosis 1	...	Diagnosis k	...	Diagnosis 1	...	Diagnosis k
New patient	●	●	●	●	●	●	●
Patient 1	●	●	●	●	●	●	●
⋮	⋮	⋮	⋮	⋮	⋮	⋮	⋮

Euclidean distance between 54-dimensional vectors would have to be computed. However, this could lead to certain problems. Two patients might be considered very similar mainly because they do definitely belong to a larger subset of diagnoses and have very similar values for these diagnoses. But they might still be different in the two or three still possible diagnoses. Therefore, we display the new patient together only with those patients in the database from the two most probable diagnoses that the classifier indicates for the new patient. Therefore, the distance between patients is computed based on 6- instead of 18-dimensional vectors. We also limit the number of patients to be displayed and show only those cases from the database being closest to the new patient.

4   Case-Centred Multidimensional Scaling

In the previous section, we have explained how the distances between the new patient and patients in the database and also distances between patients in the database can be computed. Based on these distances, MDS can be applied to obtain a visualisation that indicates how well the new patient can be classified by the computer-based decision support system. In this way, the medical doctor obtains information about where the decision support system stands and whether further information about the patient is needed to be able to distinguish well between the two closest diagnoses.

In principle, one could apply MDS directly to the corresponding distances or precalculate an MDS visualisation for the patients in the database and use the technique described in [6] to add the new patient to the visualisation. This would, however, imply that all distances are equally important: the distances between the new patient and the patients in the database in the same way as distances between patients in the database. But the focus of our visualisation is the new patient and where he should be positioned in comparison to the patients in the database.

Therefore, we introduce a case-centred version of MDS for a 3D-visualisation. The new patient is placed in the centre of the visualisation, i.e. at the origin of the coordinate system and we preserve the distance of the new patient to each of the patients in the database exactly. This is achieved by placing each patient in the database on the surface of a sphere around the origin of the coordinate system. The distance to the new patient is used as the radius of the sphere. This can be considered as a constrained MDS problem. The patients in the database cannot be positioned arbitrarily, but are constraint to the surface of their corresponding sphere.

The formalisation of this constraint MDS problem and an algorithm to solve it is presented in the following section. We use polar coordinates to simplify the problem.

Polar coordinates where already used for MDS in [7] for two-dimensional and in [8] for three-dimensional representations. However, in contrast to our proposed method, these approaches – like ordinary MDS – do not focus on a specific patient or data object in the visualisation. All patients or data objects are treated equally. But for our purposes, it is essential to primarily focus on the new patient in relation to the patients in the database and only then to consider the relations between the patients in the database.

5   Algorithm for Case-Centred Multidimensional Scaling

For each patient in the database, we have the freedom to choose the position on the surface of the sphere with the radius corresponding to the distance to the new patient. The objective function for the Sammon mapping (3) becomes for a 3D-visualisation

$$
E = \frac{1}{4} \sum_{i=1}^{n-1} \sum_{j=i+1}^{n} \frac{\left((x_i - x_j)^2 + (y_i - y_j)^2 + (z_i - z_j)^2 - d_{ij}\right)^2}{d_{ij}} \tag{4}
$$

where we assume that we have n patients in the database. The constant factor $\frac{1}{4}$ is introduced for convenience reasons as we will see later on. d_{ij} is the distance between patients i and j in the database. (x_i, y_i, z_i) are the coordinates for positioning patient i in the visualisation.

The objective function (4) should be minimised under the constraints

$$
x_i^2 + y_i^2 + z_i^2 = r_i^2 \tag{5}
$$

where r_i^2 is the distance of the new patient to patient i in the database. In this way it is guaranteed that the distances between the new patient and the patients in the database are represented without error in the visualisation.

The minimisation of the objective function (4) under the constraints (5) is based on a gradient descent method. In order to take the constraints (5) into account, we rewrite Eq. (4) in spherical coordinates. For the computation of the gradient, we need to calculate the partial derivatives with respect to all parameters. For the sake of simplicity, we only use spherical coordinates for the record i for which we want to calculate the corresponding derivatives. Then the addend in Eq. (4) becomes

$$
\frac{\left((r_i \sin(\theta_i) \cos(\varphi_i) - x_j)^2 + (r_i \sin(\theta_i) \sin(\varphi_i) - y_j)^2 + (r_i \cos(\theta_i) - z_j)^2 - d_{ij}\right)^2}{d_{ij}} \tag{6}
$$

where $0 \leq \theta_i < \pi$ and $0 \leq \varphi_i < 2\pi$.

The partial derivatives w.r.t. θ_i and φ_i are

$$
\frac{\partial E}{\partial \theta_i} = r_i \cdot \sum_{\substack{j=i \\ j \neq i}}^{n} \frac{1}{d_{ij}} \cdot
$$

$$\Big((r_i \sin(\theta_i)\cos(\varphi_i) - x_j)^2 + (r_i \sin(\theta_i)\sin(\varphi_i) - y_j)^2 + (r_i \cos(\theta_i) - z_j)^2 - d_{ij}\Big) \cdot$$

$$\Big(\quad (r_i \sin(\theta_i)\cos(\varphi_i) - x_j)\cos(\theta_i)\cos(\varphi_i)$$
$$+ (r_i \sin(\theta_i)\sin(\varphi_i) - y_j)\cos(\theta_i)\sin(\varphi_i)$$
$$- (r_i \cos(\theta_i) - z_j)\sin(\theta_i) \quad \Big) \tag{7}$$

and

$$\frac{\partial E}{\partial \varphi_i} = r_i \cdot \sum_{\substack{j=i \\ j \neq i}}^{n} \frac{1}{d_{ij}} \cdot$$

$$\Big((r_i \sin(\theta_i)\cos(\varphi_i) - x_j)^2 + (r_i \sin(\theta_i)\sin(\varphi_i) - y_j)^2 + (r_i \cos(\theta_i) - z_j)^2 - d_{ij}\Big) \cdot$$

$$\Big(\quad -(r_i \sin(\theta_i)\sin(\varphi_i) - x_j)\sin(\theta_i)\sin(\varphi_i)$$
$$+ (r_i \sin(\theta_i)\sin(\varphi_i) - y_j)\sin(\theta_i)\cos(\varphi_i) \quad \Big), \tag{8}$$

respectively.

One could first use random values for the parameters x_i and y_i ($i = 1, \ldots, n$), i.e. position each of the patients in the database randomly on the surface of his corresponding sphere. But we are more interested to make smaller mistakes between the patients in the database who are close to the new patients, i.e. those patients whose associated spheres have a radii. We insert the patients stepwise into the visualisation in ascending order with respect to their distances to the new patient (radii of their associated spheres).

By reordering the patients in the database with respect to their distance to the new patient, we can assume without loss of generality that $r_1 \leq r_2 \leq \ldots \leq r_n$ holds. We then proceed in the following way.

1. The first patient in the database, i.e. the one closest to the new patient, will be positioned at point $(x_1, y_1, z_1) = (r_1, 0, 0)$.
2. If $r_1 + \sqrt{d_{12}} \geq r_2$ and $r_1 + r_2 \geq \sqrt{d_{12}}$ hold, then the patient in the database second closest to the new patient can be positioned in such a way that the distance $\sqrt{d_{12}}$ between the first two patients in the datbase can also be represented exactly in the visualisation. In this case, choose $(x_2, y_2, z_2) = (r_2 \cos(\alpha), r_2 \sin(\alpha), 0)$ for the second patient in the database where

$$\alpha = \arccos\left(\frac{r_1^2 + r_2^2 - d_{12}}{2r_1 r_2}\right).$$

If $r_1 + \sqrt{d_{12}} < r_2$ holds, then the second patient in the database will be positioned at $(x_2, y_2, z_2) = (r_2, 0, 0)$ in order to minimise the error for the distance d_{12}.
If $r_1 + r_2 < \sqrt{d_{12}}$ holds, then the second patient in the database will be positioned at $(x_2, y_2, z_2) = (-r_2, 0, 0)$ in order to minimise the error for the distance d_{12}.

3. For positioning the third point, the problem of three intersecting spheres needs to be solved: The third point should have a distance of r_3 to the origin of the coordinate system – where the new patient is located – and distances d_{13} and d_{23} to the points representing the first and the second patient in the database. In case there is a solution to the problem of intersecting spheres, choose (x_3, y_3, z_3) where

$$x_3 = \frac{r_1^2 + r_3^2 - d_{13}}{2r_1},$$

$$y_3 = \frac{r_3^2 - d_{23} + x_2^2 + y_2^2}{2y_2} - \frac{x_2}{y_2} x_3,$$

$$z_3 = \sqrt{r_3 - x_3^2 - y_3^2}.$$

If there is no solution for the problem of three intersecting spheres, i.e. if $r_3 - x_3^2 - y_3^2 < 0$ holds, then the position of the third point is determined by a gradient method as described in the following step.

4. When the positions $(x_1, y_1, z_1), \ldots, (x_k, y_k, z_k)$ for the patients with $r_1 \leq \ldots \leq r_k \leq \ldots \leq r_n$ have been determined, the position $(x_{k+1}, y_{k+1}, z_{k+1})$ for the next patient is determined by a gradient descent method based on the gradient in terms of the variables θ_{k+1} and φ_{k+1} given in Equations (7) and (8), respectively, where $i = k + 1$. For the position $(x_{k+1}, y_{k+1}, z_{k+1})$ we only consider the patients $1, \ldots, k$.

The above described algorithm positions the points for the patients in the visualisation step by step in increasing order of their distances to the new case, in order to obtain a quick solution. Of course, one could also use this as an initialisiation and then optimise the positions again based on a similar gradient descent method, but now always taking all points into account. However, this would also lead to higher computational costs.

6   Examples

As mentioned above, we include only those patients from the database in the visualisation with the two closest diagnoses that are considered possible for the new patient. An ideal result would look like the one for the artificial data set in Figure 1 where the new patient – the red cone – is only surrounded directly by patients represented by green spheres, i.e. all patients close to the new patient have the same diagnosis. Patients with the second most probable diagnosis marked by blue cubes are all quite far away from the new patient.

Figure 2 shows the data of a 17 year old boy suffering from leukemia and pneumonia. He was initially admitted for pneumonia, but further tests revealed an acute leukemia (ALL). Interestingly, both diagnoses are valued by the system. The green spheres indicate other patients with malignant hematological diseases. The red triangle symbolizes the patient under discussion and the blue cubes stand for children with pneumonia. Of note, the doctors first treated only pneumonia in this patient and the leukemia has only

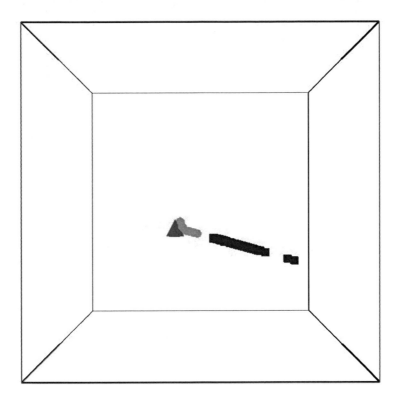

Fig. 2. A difficult patient for the doctors

been detected after transferal to a tertiary hospital. Usage of the diagnostic aid would have been helpful for treatment of this boy.

Figure 3 shows the case of patient where it was difficult to find the right diagnosis. In this 8 year old boy with high fever and headache an inflammation of the brain (meningitis) was suspected. Results of additional diagnostic tests were negative for this differential diagnosis. Two days later, he developped severe abdominal pain. Now, the doctors in charge assumed that "appendicitis" could be the cause of the problems. However, the surgeons remained sceptical and ordered a computer tomography (CT) of the abdomen. Surprisingly, the CT scan showed a large abscess formation in the kidney. To our big surprise, the computer system was already suggesting an inflammation of the urinary tract followed by systemic infection. Both differential diagnoses are apparently correct and better than those posed by the doctor on duty.

The patient in Figure 4 came to the pediatric emergency department with a short history of abdominal pain. The 11 year old boy had fever and nausea. Further tests showed a normal leukocyte number and a moderately elevated C-reactive protein. An appendicitis was suspected, but the surgeon was not convinced. Therefore, sonography was performed. Here, small amounts of fluid in the abdominal cavity were seen. As a consequence, an appendectomy was done revealing a severe inflammation of the appendix. The diagnostic tool gave strong arguments for the differential diagnosis "appendicitis" already in admission. This might help for fast track procedures.

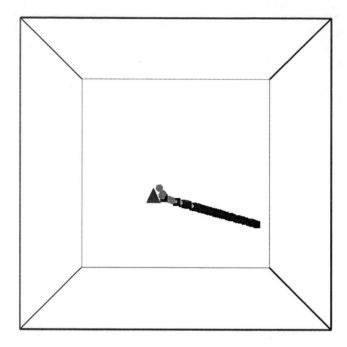

Fig. 3. A case where the system proposes the right diagnosis, but with less certainty

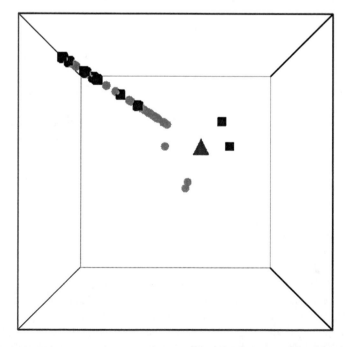

Fig. 4. Another case where the system proposes the right diagnosis, but with less certainty

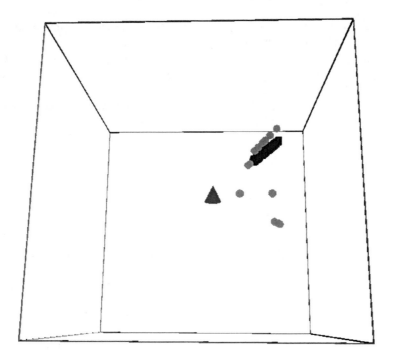

Fig. 5. Visualisation of a data set obtained from a healthy person

Figure 5 shows an interesting example with an "ideal" virtual healthy patient. The red triangle does not represent a specific patient, but is a virtual person obtained by averaging the data of all healthy persons in the database. The green spheres correspond to healthy patients in the database. Although the green spheres, representing the correct diagnosis, are closest to the virtual healthy person, they are as close to the red triangle as in the other examples. This is probably an indication for the curse of dimensionality [9,10]. The original data set contains 26 clinical and laboratory parameters, so that – according to the curse of dimensionality – the density at the centre of this 26-dimensional data set is quite low.

7   Conclusions

Wrong diagnoses carry an enormous risk for patients. Clinical decision support systems try to reduce this risk, but such applications are not widely used in daily practice. Our new diagnostic data mining tool demonstrated good results to compute a diagnosis using 26 clinical and laboratory parameters frequently used in a paediatric emergency department. To even increase the potential benefit for a user of this diagnostic support system, the display of the data is of considerable relevance. Any (tired) doctor will profit from data display immediately indicating "diagnostic clarity" or even "diagnostic vagueness". This feed-back given from the computer back to the doctor might result in an impulse to order additional diagnostic tests to provide additional security. Consequently, the risk of posing wrong diagnoses should be decreased.

Our method is not restricted to patient data, but can be applied to any type of classification problem to visualise how well a nearest neighbour classifier or a classifier ensemble can classify a specific new object.

In this paper, we have focused on 3D-visualisation. It is straightforward to derive a corresponding algorithm for a simplified 2D-visualisation.

References

1. Grigull, L., Lechner, W.: Supporting diagnostic decisions using hybrid and complementary data mining applications: a pilot study in the pediatric emergency department. Pediatric Research 71, 725–731 (2012)
2. Kruskal, J., Wish, M.: Multidimensional Scaling. SAGE Publications, Beverly Hills (1978)
3. Berthold, M., Borgelt, C., Höppner, F., Klawonn, F.: Guide to Intelligent Data Analysis: How to Intelligently Make Sense of Real Data. Springer, London (2010)
4. Shawe-Taylor, J., Cristianini, N.: Kernel Methods for Pattern Analysis. Cambridge University Press, Cambridge (2004)
5. Kuncheva, L.: Combining Pattern Classifiers: Methods and Algorithms. Wiley, Chichester (2004)
6. Pekalska, E., de Ridder, D., Duin, R., Kraaijveld, M.: A new method of generalizing Sammon mapping with application to algorithm speed-up. In: Boasson, M., Kaandorp, J., Tonino, J., Vosselman, M. (eds.) ASCI 1999: Proc. 5th Annual Conference of the Advanced School for Computing and Imaging, Delft, ASCI, pp. 221–228 (1999)
7. Rehm, F., Klawonn, F., Kruse, R.: MDS_{polar}: A new approach for dimension reduction to visualize high dimensional data. In: Famili, A.F., Kok, J.N., Peña, J.M., Siebes, A., Feelders, A. (eds.) IDA 2005. LNCS, vol. 3646, pp. 316–327. Springer, Heidelberg (2005)
8. Rehm, F., Klawonn, F.: Improving angle based mappings. In: Tang, C., Ling, C.X., Zhou, X., Cercone, N.J., Li, X. (eds.) ADMA 2008. LNCS (LNAI), vol. 5139, pp. 3–14. Springer, Heidelberg (2008)
9. François, D., Wertz, V., Verleysen, M.: The concentration of fractional distances. IEEE Trans. Knowl. Data Eng. 19(7), 873–886 (2007)
10. Jayaram, B., Klawonn, F.: Can unbounded distance measures mitigate the curse of dimensionality? Int. Journ. Data Mining, Modelling and Management 4, 361–383 (2012)

Mobile Platform for Executing Medical Business Processes and Data Collecting

Jerzy Brzeziński, Anna Kobusińska, Jacek Kobusiński,
Andrzej Stroiński, and Konrad Szałkowski

Institute of Computing Science, Poznań University of Technology
Piotrowo 2, 60-965 Poznań, Poland
{jkobusinski,akobusinska,astroinski}@cs.put.poznan.pl
jbrzezinski@put.poznan.pl, kszalkowski@gmail.com

Abstract. Medicine becomes more and more complex domain. The process from patient registration through to the provision of the right treatment becomes complex and sophisticated, and any mistakes or inaccuracies can have a significant consequences for both the patient and the care professionals. This paper presents a MMDCP — Medical Mobile Data Collecting Platform, which aims is to reduce the error rate and speed up the process of data collection. Since medical staff should have the access to medical data from any place of the healthcare facility, the MMDCP provides it's users the mobility. The proposed platform also distinguishes itself with flexibility, simplicity of maintenance and integration.

Keywords: data collecting, business process, electronic health record, integration, SOA, REST.

1 Introduction

Medicine becomes more and more complex domain. The process of patient treatment, the amount of various drugs treatments available, and other aspects of modern healthcare become increasingly sophisticated and difficult. On the other hand, the expectation of a greater efficiency means that the time spent on the execution of medical tasks is shrinking.

These aspects can lead to errors and mistakes unacceptable in any medical environment. In the process from patient registration through to the provision of the right treatment, any mistakes or inaccuracies can have significant consequences for both the patient and the care professionals.

While the training, procedural checks, double checks and clear processes can reduce error rate, the humans are still the weakest link. Care professionals are required to record information on the performed medical procedures and the medications given to patients. Unfortunately, when humans read or transcribe information, there is a possibility that this process will lead to an error. Since the process of recording medical information is relatively time consuming, it is often carried out post-factum, which only increases probability of mistake.

G. Huang et al. (Eds.): HIS 2013, LNCS 7798, pp. 149–159, 2013.

Therefore it is important to provide tools for medical staff, which will support collecting data in a way that requires less attention and time to do so. As care professionals have administrative tasks that are frequently very time-consuming, there is only a limited time available for the patient. By maximizing the efficiency of such tasks with the use of just mentioned tools, more time can be spent providing quality patient care. Moreover, the acquired data can be analyzed and utilized to automatically fill in patient electronic health record (EHR).

Another crucial feature of considered a tool is mobility. Medical staff should have an access to the medical data from any place of the healthcare facility, whether they are in the hospital pharmacy or close to the patient's bed on the ward.

According to HIMSS [17] the significant benefits can be achieved, e.g., by using barcode technology in patient registration and admission process, patient safety, clinical care delivery, tracking and accounting, product logistics management coordination. The problem with practical implementation is the real workflow and business processes that occur in healthcare units e.g. hospitals or other long term care facilities are very individual and unique. Thus the information system that supports them should be very flexible, easily configurable and transparent for the end user.

To address above mention problems, in this paper we propose a MMDCP — a mobile platform that supports automatic medical data collection. The proposed platform provides mobility, simplifies and speeds up the medical data entry process.

The paper is organized as follows. In Section 2 we describe the design and implementation of the platform. Next, in Section 3 the approach to integration of our platform with HIS systems is presented. Section 4 shows the practical application of the platform. Finally, in the last Section we conclude the paper.

2   Mobile Medical Business Process Architecture and Execution Environment

The implementation of MMDCP medical platform proposed in this paper is based on the Service Oriented Architecture (SOA) [16]. According to the SOA, the system functionality is distributed among independent applications, called web services. Such web services can be composed into so called business process when more complex and sophisticated functionality is required. In the result of SOA-based architecture of MMDCP, the functionality of the proposed platform can be easily expanded without changing already offered functions, which is important in case of all medical systems. Additionally, due to the service-oriented architecture of MMDCP, an easy and flexible integration of the proposed platform with various HIS systems is possible.

Nowadays, one of the popular approaches to implement SOA is the RPC-based approach, where Web Services and business process are implemented with a huge stack of standards like: WSDL [9], SOAP [7], BPEL [12] or set of WS-* standards (eg. [11], [10]). However, at present, also a new approach called REST [15] is getting more and more attention. In contrast to the conventional

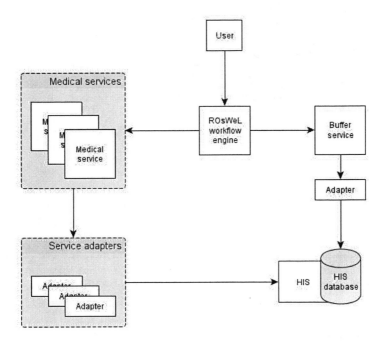

Fig. 1. Architecture of mobile medical business process platform (MMDCP)

Web Service, RESTful web services [18] are lightweight, fast and well suited to the current web architecture [8] (HTTP protocol). Additionally, RESTful web services organize functionality of the system into collection of resources available via network and provide a uniform interface to manage them. Consequently, systems implemented accordingly to REST paradigm are easier to integrate with other information systems, even those that already exist. This is a result of the use of HTTP protocol' [5] semantics which is well known and 7 used. Therefore, in MMDCP the REST paradigm was used as SOA implementation. The designed and implemented MMDCP system architecture is presented in Figure 1.

The central element of the proposed platform is ROsWeL business process execution engine [13], which is an execution environment for medical business processes. In order to create in the considered environment a medical business process the following steps are made. First the description of the business process is prepared using ROsWeL language [14] syntax. Then, the document with the defined business process is uploaded to the specified URL address of the engine service. Next, based at uploaded business process description, the new Web Service implementing the desired functionality is created and installed at ROsWeL Workflow Engine. In consequent, the business process in a form of a RESTful Web Service is provided to the user. Such a RESTful Web Service can be invoked on almost any type of device (personal computer, smart phone or tablet). In order to fully utilize the proposed platform, such a device has only to support HTTP communication protocol, and HTML format to display a user interface. In practice, it can be any device that has a web browser installed.

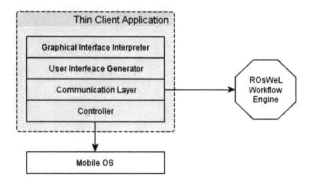

Fig. 2. "Thin client" architecture

It is important to note, that ROsWeL engine is responsible for processing both logic, and communication between the system components, and the end-user device, including the information on how to display data, what information should be enter to the system, and how. In other words, ROsWeL engine allows to compose several advanced medical services (being a collection of RESTful medical resources) providing a complex functionality, by invoking single medical services in a defined in a ROsWeL file order. Details on the description of a business process can be found in [14] and information on design and implementation of the business process engine are in [13].

Second element of the proposed MMDCP platform is the end-user universal mobile application. Its main task is to initialize business process execution and the active participation in it. The important feature of such a mobile application is its ability to generate the user interface based on the data transmitted from the ROsWeL engine during the business process execution (called "on-the-fly"). Such an interface allows medical staff to enter the data either using a keyboard, a camera or a dedicated laser reader integrated with the device or connected as a peripheral device. In the current state of a development of MMDCP, its user can enter several types of data: strings, images and barcodes. Using the built-in camera or dedicated barcode scanner there is possibility to scan either 1D bar codes (the cheapest and widely used), or 2D bar codes (more expensive but much more reliable) like QR-code, Aztec etc. In the consequence, we will call the mobile application a "thin client", which means that even a very simple and inexpensive device can play such a role within the proposed platform.

Up to date, a "thin client" is available for the following platform: Windows Mobile (v6.1, v6.5), Windows Phone, Android and PC with a web browser. It is important to add that regardless of the device and its quality, the graphical interface design of the proposed application looks almost the same. This has been achieved by our thin client architecture depicted in Figure 2.

The architecture of mobile application is modular and consists of the following modules: Controller, Communication Layer, User Interface Generator and Graphical Interface Interpreter.

The Controller module is used to invoke native functions of different mobile devices. It is used for setting or getting the mobile device settings like language, sound, connection and uses them during a business process execution. On the top of the Controller module the Communication Layer is built. Its main goal is to send and retrieve messages between ROsWeL engine and a thin client application. Above the Communication module is the User Interface Generator placed. It retrieves messages received by Communication Layer in order to generate optimal interface for a device (screen resolution etc.). The interface's data are described with a small subset of HTML5 tags, enhanced with some missing functionality, such as a barcode and some optimizations, which consist of postponement the generation of the part of the user interface in order to accelerate their display on certain potentially weak devices. The missing part of interface is added to the already generated interface later, on the user's demand. The last module of mobile application is a Graphic Interface Interpreter, which simply draws generated interface on the mobile device screen.

3   The Integration of MMDCP with Hospital Information Systems

Nowadays there are many Hospital Informations Systems (HIS) on the market. They are required to meet demands of the fast data access, high customizability, and stringent safety. Such requirements result in complex architecture of such systems, and their custom software solutions. Therefore, the integration with HIS systems is a very difficult task. The MMDCP platform provides an internal mechanisms to integrate it with the existing HIS systems.

In order to achieve maximal interoperability, the proposed MMDCP does not communicate with a Hospital Information System (HIS) in a direct manner. In case of reading medical data, ROsWeL invokes medical services (resources), which are network interfaces for data stored in the HIS. Additionally, to allow an easy integration with various HIS systems, medical services use so called adapters to adjust the inner HIS data format (e.g. database structure) to the one understood by the MMDCP platform. The details of the way of exchanging the format are discussed in the following part of this Section. The last element of the platform is a buffer service. It is used in order enable recording some data in HIS database. Since HIS systems have various architectures and functionality, and require a strict control of what can be recorded, the proposed platform adds a buffer element to implement so called "delayed write". The main idea of a delayed write is to record temporarily data into buffer instead of HIS database directly. Afterwards, the HIS system retrieves on demand data stored in a buffer via buffer adapter and updates its medical records in internal database. This approach allows to control the data that will be stored in HIS database, and provides a high level of interoperability.

The proposed solution consists of two Java web services utilizing Jersey [6] technology and using Hibernate [3] ORM mapping framework. First of them is a Universal Identification Services (UIS), which task is to export information from

HIS database to the business process engine in order to provide data objects required by business processes. This service after a few simple configuration steps creates REST interface for accessing objects using different data types, identifying them by scanned barcodes or their natural database ids.

The second web service — Universal Buffer Service (UBS) — is burdened with task of gathering in an universal way the information along with multiple useful meta information, collected during the business processes execution.

Due to the fact that each of the above mentioned services performs different tasks, UIS and UBS are presented in the following sections.

3.1   Universal Identification Service

Primary task of the UIS (architecture in Figure 3) is to fetch information from the HIS system and present it to the business process engine, which can therefore use the obtained data to complete its tasks. The proposed service neither inter-feres with existing HIS infrastructure, nor requires to create redundant data structures. Its requirements are very strict and simple, and are the following. Database system must cooperate with Hibernate (either by standard terms or by specific dialect and driver). Additionally, each object in that database should be supplied with barcode persisted in a string format. Moreover, the type of data stored in the table's columns must be Hibernate's standard basic types [4] (this last requirement is planned to be uplifted in the future work). As the result, the UIS will support also other types of columns: references to another tables, complex types, collection types.

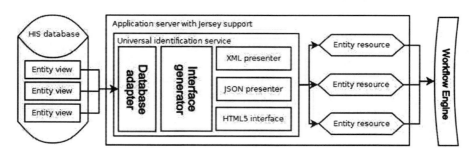

Fig. 3. Universal Identification Service architecture

Configuration process requires from the platform administrator to take a few simple steps. The Administrator should first consider what is required in the business process. Then he/she should prepare database's tables and views, and the configuration storage place. Then should be filled out the database configu-ration file and created Hibernate mappings for the views.

Since to complete a business process users need some data, this data can be presented to business process engine by the UIS. Choosing the right data is crucial, as too much information may be confusing, and too few will prevent business process user from completing his task successfully. When the administrator knows what data is required, a database presentation should be prepared, because in the HIS systems data is stored in different tables or views. The UIS requires that each object class should be stored in one table or view with using basic types such as string or integer. Often this step is negligible, because data is already presented in the right way. Therefore, the administrator should prepare storage place for configuration files, a separate folder with adequate security rights if it should not be read by applications other than application server. The UIS searches for configuration files in the preconfigured location set in Java's `System.getProperties()` facility. In Tomcat this values can be found in `catalina.conf`facility. This folder should contain Hibernate database connection configuration stored in `hibernate.cfg.xml` file - a XML document. Also the application server on which the UIS is installed should be equipped with the proper JDBC connection drivers mentioned in the file. The final step for platform administrator is to create the hibernate mappings for desired objects. Those mappings must be supplemented by only one mandatory Hibernate meta-argument "barcode". Those meta-argument should mark mentioned, mandatory textual field containing the barcode.

Universal Identification Service on the basis of configuration files creates the REST interface. This interface is available instantly after the service's start under the service's URL address configured by the application server administrator. Each of the configured object classes can be accessed as separate resource. Such a resource informs also the user about the object's properties. Objects of given class can be accessed by identifying them with their natural database id or by the barcode. Each of the mentioned resources can be presented by the UIS in three different formats, namely: XML, JSON, HTML5. Whether the first two are common standard of web application communication formats, the last one is meant for humans.

3.2   Business Process Buffer Service

Business processes are often created with a purpose to perform some process along with the user, and during that process data is collected. This data can contain useful information for the HIS system, for example a number of syringes that remain in warehouse or amount of drug injected. Such a data should be stored somehow. This problem may be solved with the usage of the Universal Buffer Service (architecture at Figure 4) — a service which delays the storage of data to the moment when the HIS system will be ready to accept it ("delayed write"). The role of the service is to collect useful meta-information. The collected information can be utilized to optimize business processes, provide additional security or perform monitoring services. Such information comprises among the others: user, who executed process, device, on which process was executed, process which was executed, host name address, execution time and application name.

Configuration process of UBS is composed of three steps: finding and creating database according to the supplied schemas, providing Hibernate database configuration file, and adding eventual security restrictions on business process

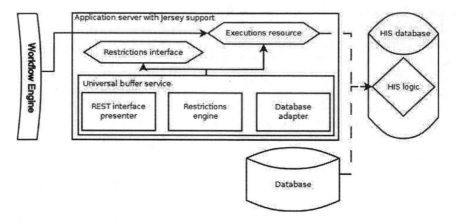

Fig. 4. Universal Buffer Service architecture

types, users or devices on which they are performed. Database, which ought to be prepared, must contain tables mentioned in the service's documentation and must cooperate with Hibernate. Database configuration file again must be put in the folder, which therefore will be read through the Java's properties facility. The last and optional step allows to restrict business process executions which will be accepted — only specific processes, users or devices can be allowed to store data inside the UBS. Also simple meta-description rules can be provided to validate the data stored by business process users in order to achieve consistency (similarly to the XSD).

Data gathered by the Universal Buffer Service is stored inside the database. It can be therefore collected by the HIS system in order to allow it to maintain its internal procedures and structures. Business process execution's data database stores object of execution, which has a mentioned meta-information. All other data is stored as its properties in a textual way, as provided by the workflow engine. This data can be accessed through database connection or through the REST interface, which provides it as a resource. REST interface also contains separate resource for configuring the mentioned restrictions. UBS presents to user additional web-based interface which allows to browse database and store additional restrictions.

Both backend services provide well defined, simple and robust interface for both the business process workflow engine (that seamlessly integrates with services), and hospital information system. Despite need of some effort to plug the Universal Buffer Service and Business Process Buffer Service, the services do not interfere with existing infrastructure. Therefore they can be integrated with almost every platform currently existing in the market.

4   MMDCP Data Collecting Process

In order to illustrate the possible usage of the proposed MMDCP platform in the healthcare facilities, an exemplary process is discussed (5). In the considered

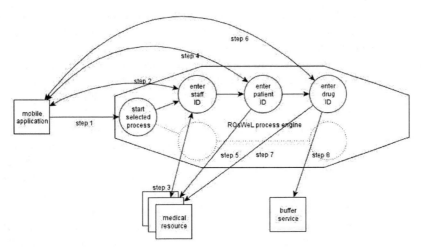

Fig. 5. Drug administration process

process the information on drugs administered to patients during medical procedures is being collected. First, medical staff using a mobile application installed at a mobile device connects to the ROsWeL engine and asks for a list of available processes. After selecting the drug administration process, ROsWeL engine executes first step of the process (step 1), which requires to enter medical staff member ID in order to verify her/his identity (step 2). ID number can be entered manually (keyboard), or scanned from the barcode placed at the ID of the employee. Next, ROsWeL engine via a Personel Identification Resource (one of the available medical resources) queries HIS database for the verification (step 3). If such a person exists in the database and is authorized to administrate drugs, the information is send back to the ROsWeL engine to proceed, otherwise authorization error is send back to the thin client. In the next step the patient's identification number same as medical staff person ID (step 4) has to be entered. The patient information is verified, and (step 5) send back by ROsWeL engine to the client in order to generate the appropriate interface at mobile device, consistent with the definition of the business process. If the above steps pass successfully, the ID number of administrated drug can be entered into the system (step 6, step 7 and step 8). It is important that usually each patient takes more then one drug. To administrate a new drug fast and easily, the MMDCP platform (ROsWeL engine) remembers medical staff and patients, and do not require to repeat first two steps of the above described process. Changing a patient, or a medical staff who takes part in the considered process, one just needs to enter a new ID of respectively patient or a medical staff.. Finally, all complete triples in form of: `<staff_id, patient_id, drug_id>` are recorded in a buffer (step 8).

5   Conclusions and Future Work

The MMDCP platform provides a flexible and easy to integrate environment that supports medical staff in the daily duties of collecting medical data. This solution has been successfully integrated with the Eskulap Hospital Information System, which is the third most used system in polish hospitals [1]. Due to the applied service-oriented architecture and technology, the integration process was completed without any major problems, and its validity has been confirmed in practice. By providing by MMDCP a possibility of defining a new business processes, it is possible to obtain a completely new functionality that enriches the Eskulap HIS system almost out of the box.

Further development of the proposed platform will be simultaneously focused on two directions. The first one will focus on extending the possible range of MMDCP practical applications. We are going to enhance the presented platform by adding a possibility to use information on the patient's insurance and the quality of that insurance in order to propose an appropriate drug or its equivalent. This functionality also makes it possible to facilitate the easier management of hospital pharmacy in case of the lack of certain drugs. The latter option concerns with the development of the ROsWeL Workflow Engine functionality. Currently, we work on adding the semantic information on the business process steps and on resources used during the process composition. Such a semantic extension will allow, e.g. to more accurately propose equivalent drugs or even equivalent resources in case of system failures.

Finally, one might consider using the HL7 [2] standard as an internal representation of medical data structure. This standard is well recognized in the medical environment and would allow to not only add new data sources to the platform but also allow to exchange messages with other systems more easily.

References

1. ESKULAP Hospital Information System, http://www.systemeskulap.pl/
2. Health Level Seven International, http://www.hl7.org/
3. Hibernate Object-Relational Mapping Framework, http://www.hibernate.org/
4. Hibernate Reference Manual, http://docs.jboss.org/hibernate/orm/4.1/
5. Hypertext Transfer Protocol – HTTP/1.1, http://www.w3.org/Protocols/rfc2616/
6. Jersey JAX-RS (JSR 311) Reference Implementation, http://jersey.java.net/
7. Simple Object Access Protocol (SOAP) 1.1, http://www.w3.org/TR/soap/
8. W3C: Web Architecture, http://www.w3.org/standards/webarch/
9. Web Services Description Language (WSDL) 2.0, http://www.w3.org/TR/wsdl20/
10. Web services addressing, ws-addressing (2004),
    http://www.w3.org/Submission/2004/SUBM-ws-addressing-20040810/
11. Web services security: Soap message security 1.1, ws-security (2006),
    https://www.oasis-open.org/committees/download.php/
    16790/wss-v1.1-spec-os-SOAPMessageSecurity.pdf
12. Web services business process execution language version 2.0 (2007),
    http://docs.oasis-open.org/wsbpel/2.0/wsbpel-v2.0.pdf
13. Brzeziński, J., Danilecki, A., Flotyński, J., Kobusińska, A., Stroiński, A.: Workflow Engine Supporting RESTful Web Services. In: Nguyen, N.T., Kim, C.-G., Janiak, A. (eds.) ACIIDS 2011, Part I. LNCS, vol. 6591, pp. 377–385. Springer, Heidelberg (2011)

14. Brzeziński, J., Danilecki, A., Flotyński, J., Kobusińska, A., Stroiński, A.: ROsWeL Workflow Language: A Declarative, Resource-oriented Approach. New Generation Computing 30(2-3), 141–163 (2012)
15. Fielding, R.T.: Architectural Styles and the Design of Network-based Software Architectures. Ph.D. thesis, University of California, Irvine (2000)
16. Krafzig, D., Banke, K., Slama, D.: Enterprise SOA: Service-Oriented Architecture Best Practices. Prentice Hall PTR, Upper Saddle River (2004)
17. Menola, F., Miller, E.: Implementation Guide for the Use for Bar Code Technology in Healthcare (2003), http://www.himss.org/
18. Richardson, L., Ruby, S.: RESTful Web Services (2007)

Prediction of Assistive Technology Adoption for People with Dementia

Shuai Zhang[1], Sally McClean[1], Chris Nugent[2], Sonja O'Neill[2], Mark Donnelly[2], Leo Galway[2], Bryan Scotney[1], and Ian Cleland[2]

[1] School of Computing and Information Engineering, University of Ulster, Coleraine Campus, Cromore Road, Co. Londonderry, BT52 1SA, Northern Ireland
[2] School of Computing and Mathematics, University of Ulster, Jordanstown Campus, Shore Road, Newtownabbey, Co. Antrim, BT37 0QB, Northern Ireland
{s.zhang,si.mcclean,cd.nugent,s.oneill,mp.donnelly,l.galway,
bw.scotney,i.cleland}@ulster.ac.uk

Abstract. Assistive technology can enhance the level of independence of people with dementia thereby increasing the possibility of remaining in their own homes. It is important that suitable technologies are selected for people with dementia, due to their reluctant to change. In our work, a predictive model has been developed for technology adoption of a Mobile Phone-based Video Streaming solution developed for people with dementia, taking account of individual characteristics. Relevant features for technology adoption were identified and highlighted. A decision tree was then trained based on these features using Quinlan's C4.5 algorithm. For the evaluation, repeated cross-validation was performed. Results are promising and comparable with those achieved using a logistic regression model. Statistical tests show no significant difference between the performance of a decision tree model and a logistic regression model ($p=0.894$). Also, the decision tree demonstrates graphically the decision making process with transparency, which is a desirable feature within healthcare based applications. In addition, the decision tree provides ease of use and interpretation and hence is easier for healthcare professionals to understand and to use both appropriately and confidently.

Keywords: Technology adoption, Decision tree, Dementia, Assistive technology.

1   Introduction

A common pathway for people with dementia (PwD), taking into consideration their increased care requirements, is to eventually move from their home environment into a care facility. One possible approach to improve this situation and to allow persons to remain living at home for longer, with lower costs and with improved levels of independence, is through the use of assistive technology. Nevertheless, PwD are generally reluctant to change their routine and in addition, are often afraid of making mistakes using technology or are simply unable to use it. It is therefore important to identify

G. Huang et al. (Eds.): HIS 2013, LNCS 7798, pp. 160–171, 2013.

the best suitable technology for the PwD - caregiver dyad. A key element in achieving this is being able to highlight which characteristics may determine the likelihood of appropriate adoption for a particular technology.

The aim of the current study is to identify the features that may influence technology adoption and based on these features develop a predictive model which can be used to determine whether a PwD may or may not adopt assistive technology. The usage of such a screening process would be of benefit to clinicians and healthcare professionals at the point of introducing a new form of assistive technology into the daily life of the PwD (e.g. in a memory clinic setting). During a patient assessment, information could be entered into the predictive model and the healthcare professional could be informed immediately as to the likelihood of success if technology were to be introduced.

A number of attempts to address the notion of prediction have been reported in the literature, for example, the Psychosocial Impact of Assistive Devices Scale (PIADS) or the technology acceptance model (TAM) [1-2]. TAM is based on the theory of reasoned action and states that the behaviour intention is influenced by perceived usefulness and perceived ease of use. This has been shown to have a direct effect on the actual behaviour [1]. A common approach when considering likely features to adoption is to separate the features into external environmental features such as social structures, the regulatory environment and infrastructure in addition to internal personal features such as utility perception, expectations and self-esteem [2]. The PIADS scale is an extension to TAM which is focused on personal features and acknowledges the existence of external features such as social networks and a larger society around which may have an impact on usage and on self-image. The PIADS scale is targeted specifically towards assistive technology. The assessment of embarrassment and other negative connotations have been included in the PIADS scale to assess the psychosocial impact of assistive technology on three levels: competence, adaptability and self-esteem. This 26-item scale, however, requires that a person is able to reflect and provide feedback on their perceptions; for PwD such a reflection may be difficult. Both the TAM and PIADS models have been criticised in the literature due to their questionable heuristic value and lack of explanatory and predictive power [3].

The organisation of this paper is as follows: Section 2 provides a synopsis of experimental setup and data collection. Section 3 describes data pre-processing, including the process of feature selection and discretisation. Section 4 presents a decision tree trained based on data with selected features using Quinlan's C4.5 algorithm. This is followed by the model evaluation, discussion and its use in the form of a set of rule-based classifiers in Section 5. Conclusions and future work is provided in Section 6.

2   Data Collection

Our Mobile Phone-based Video Streaming (MPVS) solution has been developed to provide reminders for everyday tasks for PwD. The system works through the delivery of video based reminders using a mobile phone [4]. The system is comprised of three components: a mobile phone based component, which has been modified to support easy interaction for PwDs and is used to deliver personalised video messages to provide reminding prompts. Upon receiving reminders, users are required to press a large button on the device, which acknowledges receipt of the reminder and causes playback of a pre-recorded video. A second element of the system provides caregivers with a touch screen and an associated software application, to record video reminders and to schedule these appropriately. The third component of the system is a server, which manages the storage, communication and transmission of data between the caregiver application and the mobile phone based application.

The MPVS system has been evaluated and improved iteratively. Throughout this iterative process a range of evaluations has been undertaken with a cohort of 40 PwDs. The technological platform was then updated in accordance with feedback from these evaluations [4-7]. In an attempt to identify the relevant features pertaining to adoption or abandonment of the system, interviews were performed with the various members of the research and development team which included: biomedical engineers, computer scientists, research nurses and geriatric consultants. Together, they identified a range of features potentially relevant to the adoption of the MPVS solution, such as appropriate infrastructure, the cognitive and physical ability to manage the system, previous experience with technology and the perceived utility of the assistive technology. In addition, the role of the carer was identified as being important, with features such as carer burden, encouragement for the PwD, perception of utility and their technology experience being identified. Following these interviews and discussions, an influence diagram was created, indicating which parameters may have an impact on which other parameters with respect to the PwD and also to the family carer in relation to adoption of the technology. Once established, the influence diagram was presented and discussed and subsequently refined in a workshop with formal carers of PwD (n=8).

Consequently, the data from the aforementioned evaluations, in line with the features identified through the influence diagram, from the 40 participants were collated from databases and patient visit logs. The features identified are presented in Table 1. The level of adoption was described by a 4-item Likert scale recorded by the research nurse, based on previously recorded notes, which indicated whether the dyad had dropped out, was non-compliant or was compliant or even eager to keep the technology. Later this scale was contracted to two classes of *Non-adopter* and *Adopter* where the *Non-Adopter class* includes the users in the categories of 'dropped out' and 'non-compliant' and *Adopter* class contains the users in the categories of 'compliant' and 'eager to keep the technology'.

Table 1. Features collected from patients through interviews and questionnaires

Feature	Data Type	Range / Options + Description
Age	Scale	20-100 (years old)
Gender	category	Male / female
Mini Mental State Examination (MMSE)	Scale	30-point questionnaire test (0-30); to screen for cognitive impairment
Previous Profession	category	Administrative Position, Caring Profession, Technician / Engineer, Other
Technology Experience Patient	category	score of 0-4, calculated from technology questions, where 0 – no mobile phone, no PC; 1 – use mobile to receive calls only, no PC or no mobile, have PC; 2 – use mobile to receive and initiate calls, no PC or use mobile to receive calls only, have PC; 3 – use mobile for initiate, receive calls and sms, no PC; use mobile to receive and initiate calls, have PC; 4 – use mobile for initiate, receive calls and sms, have PC
Broadband	category	Yes / No
Mobile_Reception	category	Issues / good
Carer_Involvement	category	Yes / No
Living_Arrangement	category	Living alone / living with somebody
Extra Support	category	Yes/No Whether the patient has care from family support or additional informal carers
Physical Health	scale	Disability Assessment For Dementia (DAD) or Instrumental activities of daily living (IADL) – independent questionnaire
If there is a carer		
Age Carer	scale	20-100 (years old)
Gender of Carer	category	Male / female
Previous Profession Carer	category	Administrative Position, Caring Profession, Technician / Engineer, Other
Health of Carer	string	indirect through issues
Completed after evaluation		
Adoption	category	Yes / No

3   Data Pre-processing

In the first instance, in the process of feature selection, the most relevant combination of features to the output *Adoption* class was selected. Following this process, the continuous features in the selected feature set were discretised as this is required for the decision tree approach.

3.1   Feature Selection

In the collected data, 32.5% of patients did not have a carer. Given the small sample size, information related to the carers, namely the carer's age, gender, previous profession and health, was therefore omitted in the data analysis. Nevertheless, such information is deemed to be potentially informative and may be considered if a larger dataset were to be made available in the future.

Feature selection was performed in a data pre-processing stage to select the most relevant features for the output class, *Adoption*. In the first instance, a pairwise significance test was performed on an individual feature against the output class to select the directly relevant features. The Chi-square or Fisher's exact test [8] was used for features of categorical type and the Mann-Whitney test was used for continuous data type features. All tests were performed using IBM® SPSS® Statistics 19.0.0. For the Chi-square tests, if the cell expected frequency fell below 5 in more than 25% of the cells of the contingency table, Fisher's exact test was used. A conventional *p*-value of 0.05 was used for the threshold of significance. Results from the pairwise significance test in the pre-processing revealed four significantly relevant features for the outcome feature *Adoption* namely *MMSE*, *Age*, *Living_Arrangement* and *Broadband*. Principle component analysis was subsequently used to identify indirectly relevant and related features, where the combination of features may have a significant relationship with the output feature *Adoption*. Three additional features were subsequently included in the reduced feature set namely *Carer_Involvement*, *Gender* and *Mobile_Reception*.

3.2   Feature Discretisation

Following the process of feature selection, the continuous features of *Age* and *MMSE* were discretised. The discretisation of the MMSE score follows the guidelines from the Alzheimer Society [9] which stated that: MMSE scores of 27 or above (out of 30) are considered normal, mild Alzheimer's disease equates to an MMSE score of 21-26, moderate Alzheimer's disease equates to an MMSE score between 10-20 and severe Alzheimer's disease equates to an MMSE score less than 10. Given the inclusion criterion of recruitment of PwDs with an MMSE score over 18 in the studies conducted, there was only a small sample size in the category of 'moderate' and none within the category 'Severe'. The distribution of Age generally follows a normal distribution as presented in Figure 1 with a mean value of 72.5 years. Based on the age histogram and to avoid possible over-fitting based on the small sample size, the discretisation was based on a division using two values 65 and 75 years.

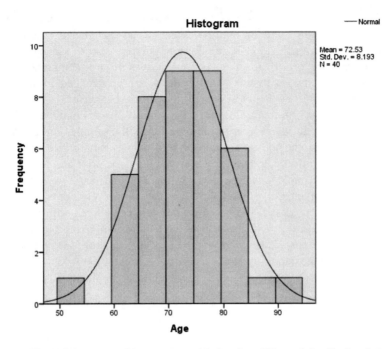

Fig. 1. Histogram of feature *Age* with the plot of Normal distribution fitting

The categorical features following the process of feature selection are *Gender*, *Broadband*= {'Mobile broadband connection', 'Fixed line broadband connection'}, *Mobile_Reception*= {'Issues', 'No issues'}, *Carer_Involvement*= {'No carer', 'Carer involved'} and *Living_Arrangement*= {'Living alone', 'Living with somebody'}. The discretisation of the continuous features *Age* and *MMSE* is presented in Table 2.

Table 2. Discretisation for *Age* and *MMSE* features selected following feature selection

	1	2	3
Age	≤65	66-75	>75
MMSE	≤20	21-26	>26

4   Classification

A decision tree (DT) is a popular data mining approach based on a top-down divide-and-conquer induction strategy [10]. The DT can be represented in a graphical tree, where leaves of a tree represent the class labels and branches represent conjunctions of features that lead to those classes. The principle of the DT is to partition the data space spanned by the input features to maximise a score of class purity based on information theory that the majority of points in a node of a tree belong to one class [11]. At each stage an attribute can be split based on the best attribute that separates the classes of the instances of that attribute. Each node can then be split into more branches. The process can be repeated a number of times. DTs are useful in healthcare related applications given that the decision making process is transparent and can be presented on graphical DTs. In addition, DTs are easy to use and interpret and they can be transformed to a set of rules which are comprehensive to non-technical decision makers to use. Such features are appreciated by healthcare professionals especially if they do not have a computational background.

For the given problem, Quinlan's C4.5 DT algorithm [12] was employed using the concept of information entropy. The criterion of splitting the samples for a node was based on an attribute with the highest normalised information gain, where information gain is the difference in information entropy between the node prior and post splitting using the attribute. The implementation was via the J48 tree classifier which is an open source Java implementation of the C4.5 algorithm in the Weka data mining tool [13]. The derived tree (noted as DT7) based on the data considered within the given study is presented in Figure 2 with the selected 7 features following the process of feature selection. In Figure 2, the light grey nodes marked *NA* represent the Non-Adopter class and the dark grey nodes marked *A* represent the Adopter class.

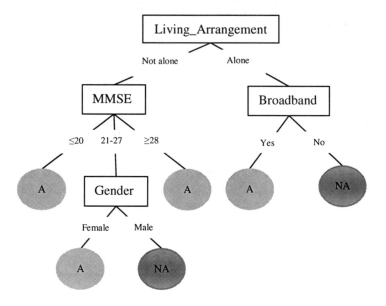

Fig. 2. The DT model DT$_7$ for technology adoption, where the light grey nodes marked *NA* represents *Non-Adopter* class and the dark grey nodes   marked *A* represents *Adopter* class.

5   Evaluation Results

The performance of the trained DT model was evaluated and compared with the model generated using a logistic regression approach. Rules were then naturally generated from the DT. The advantages of the DTs are further discussed, especially in the context of healthcare related applications.

5.1   Evaluation

For the purposes of model evaluation, in situations where there is not a sufficient amount of data available, cross-validation can be used to measure the error rate of a learning scheme [14]. The evaluation of the DT_7 model performance within the current study was performed using 4-fold cross validation. To assist in obtaining reliable results, the 4-fold cross validation was repeated ten times.

Both the training and test data should be representative samples of the underlying problem. To meet this criterion, the order of the data was initially randomised. For the training and test data sampling, a stratification process was applied to ensure that each class was properly represented in both the training and test datasets for each fold. Based on this approach, at each cross-validation stage, the dataset (n=40) was split into 4 equal proportions with assigned folds (n=10). Each fold of data was in turn used as test data with the remainder being used for training. Therefore, the model was trained on 3 folds of the data which equated to 75% of the entire samples and subsequently tested with the remaining 25%. Given that it was a 4-fold cross validation, the training and testing process was repeated 4 times on a different training set and test set, where every instance was used exactly once for testing. The 4 performances from each fold were averaged to yield a classification accuracy value. The cross-validation process was repeated 10 times with different stratified random sampling in an effort to produce reliable results. Finally, the model performance was averaged over the 10 cross-validation repetitions with the standard deviation, also presented to demonstrate variability.

The performance results are presented in Table 3 for the evaluations on the training data and the 10 iterations of the 4-fold cross-validation. The corresponding standard deviation values are presented alongside each mean value in brackets.

Given the small amount of data available, the performance of the models in the cross-validation process built from 75% of the data is likely to be a pessimistic estimation of the true model. To compensate for this, the ultimate model performance is combined with the resubstitution performance on the training data which provides an optimistic estimate of the model performance on the new data [14]. The estimate model prediction performance is thus estimated as follows:

$$Acc = 0.75 \cdot Acc_{test} + 0.25 \cdot Acc_{train}$$

The final estimated performance, combining both performance on the training and test data, is presented in the last column of Table 3.

In our previous work, a logistic regression approach [15-16] was employed to create a decision making model based on binary features from the same data collection. For comparison with the DT approach, a logistic regression model was generated (represented as model LR_7) using the classifier class in the Weka tool [13]. Similarly to the evaluation for the DT model, the evaluation of the logistic regression model on the data was performed using 10-time repeated 4-fold cross-validation. The comparison of the performance between DT_7 and LR_7 is presented in Table 3.

Table 3. Prediction accuracies in percentage (%) for the DT model and a logistic regression model (standard deviation figures are presented in brackets)

Models	Training data (%) Acc_{train}	Average of 10 times 4-fold cross-validations (%) Acc_{test}	Estimated performance (%) Acc
DT$_7$	90	72 (2.84)	76.5
LR$_7$	85	72.5 (7.73)	75.625

The evaluation results show comparable performance between the DT and the logistic regression based approach. A pair-wise t-test results ($p=0.894$) showed no significant difference between the performances of the two models DT$_7$ and LR$_7$. Nevertheless, the DT model has the ability to demonstrate graphically how the model works, a desirable feature within healthcare based applications. The DT model demonstrated ease of use and interpretation, hence is easy for healthcare professionals to understand and to use more appropriately and confidently. Additionally, a DT can be easily transformed to a set of rules which is comprehensive to non-technical decision makers/users to employ and understand.

5.2   Rules for Decision Making

Potential decision makers of the MPVS system can easily follow the information of the DT from the top down until reaching a class point (a leaf in the tree) where the answer will be provided. Based on Figure 2 the first decision to be taken relates to whether the potential user is living alone. If, for example, the person is living alone, then a check should be made if he/she has broadband at home or not. If the person has broadband at home, then the person can be considered as a potential adopter of the MPVS solution. If, however, he/she does not have broadband, then the person is considered to be a potential non-adopter.

A set of rules were extracted from the tree presented in Figure 2 for a more comprehensive decision making process. Rules extracted from the tree are presented in the following with their support s and confidence c values in the brackets following that rule. Support s of a rule measures the proportion of data that contains the specified values of the features in the antecedent of the rule. Confidence c of a rule measures, among the data with antecedent of the rule, the proportion of the data with the predicted class in the consequent of the rule [14]. Support reflects the amount of information available to build a rule and confidence reveals the reliability of the rule used for classification.

- <u>Rule 1</u>: If (Person is living alone) AND (Person has a broadband connection), then the Person is of an *Adaptor* class ($s= 0.05$, $c= 1.0$);

- <u>Rule 2</u>: If (Person is living alone) AND (Person does NOT have a broadband connection), then the Person is of a *Non-Adaptor* class   (s= 0.175, c= 0.857);
- <u>Rule 3</u>: If (Person is living with somebody) AND (Person's MMSE is in category 1), then the Person is of an *Adaptor* class   (s= 0.05 , c= 0.5);
- <u>Rule 4</u>: If (Person is living with somebody) AND (Person's MMSE is in category 3), then the Person is of an *Adaptor* class   (s= 0.425, c= 0.941);
- <u>Rule 5</u>: If (Person is living with somebody) AND (Person's MMSE is in category 2) AND (Person is Female), then the Person is of an *Adaptor* class   (s= 0.175, c= 1.0);
- <u>Rule 6</u>: If (Person is living with somebody) AND (Person's MMSE is in category 2) AND (Person is Male), then the Person is of a *Non-Adaptor* class (s= 0.125, c= 0.8).

Rule 3 in the above rule set is used for prediction of MPVS adoption for people who have moderate memory impairment and are living with somebody. This rule has a particularly low confidence of 50%. The decision of whether the patient will adopt the MPVS solution is unclear with no strong evidence of the person being either an adopter or a non-adopter. However, the support s of the rule is as low as 0.05. It is not surprising that the rule confidence is low here as there is not sufficient information to build this rule for high prediction performance. Additional features may need to be investigated to support a more confident decision or additional data are required.

There are two rules for the non-adopter class: Rule 2 and Rule 6. If a person is living alone and does not have a broadband connection, the person is a non-adopter for all categories of memory impairment. If a male patient with mild memory impairment is living with somebody, then they are a non-adopter of the MPVS system.

6   Conclusions and Future Work

People with Dementia are often reluctant to accept changes to their routine. When presented with assistive technology, such as the MPVS solution, it is very important to take into account an individual user's characteristics and needs and subsequently make an appropriate decision on their probability of adoption. If there is a failure to recognise a potential adopter, a significant opportunity is missed in terms of the potential impact using an assistive technology that may have allowed them to stay in their home for longer. On the other hand, if a non-adoption solution is inappropriately prescribed, not only will there be financial implications, the failure to interact with the device can affect the mood of the PwD and subsequently have a negative impact on their quality of life and the quality of life of their care giver. In this paper, we introduced a DT approach with the aim of predicting the likely adoption or otherwise of the MPVS solution for PwD. The performance of the DT was comparable with a logistical regression model. The advantage of DT models is that they are easy to use and interpret from a non-technical user's perspective. Rules can easily be extracted from DT models. Their simplicity makes them usable by any decision makers or relevant health professionals who do not have a technical or computational background.

Though our prediction model produces encouraging and promising results, the study is limited by the small sample size. An improved model can be built with additional data available and/or additional features; we plan to take this into account in our future work. A collaborative project with Utah State University has recently been established (Technology Adoption and Prediction Tool for Everyday Technologies), which will allow this work to be extended to a larger sample size. This collaboration will make use of information from the Cache County Study on Memory in Aging (CCSMA) and the Utah population database.

Additionally, there is an imbalance in the distribution of the two classes. The performance is biased towards the majority class. The evaluation of our DT_7 model showed the type I error (false positive) of 7% for adopters, which is lower than the type II error (false negative) of 17% for non-adopters. Consideration is therefore required to improve the algorithm in relation to this issue. The cost of two types of classification errors can be built into the classification model in the future to minimise the total cost of misclassification.

Acknowledgements. The authors wish to acknowledge support from the EPSRC through the MATCH programme (EP/F063822/1 and EP/G012393/1). The views expressed are those of the authors alone.

References

1. Yen, D.C., Wu, C., Cheng, F., Huang, Y.: Determinants of users' intention to adopt wireless technology: An empirical study by integrating TTF with TAM. Computers in Human Behavior 26, 906–915 (2010)
2. Scherer, M., Jutai, J.J., Fuhrer, M., Demers, L., Deruyter, F.: A framework for modelling the selection of assistive technology devices (ATDs). Disability and Rehabilitation: Assistive Technology 2, 1–8 (2007)
3. Chuttur, M.: Overview of the Technology Acceptance Model: Origins, Developments and Future Directions. Sprouts: Working Papers on Information Systems 9(37) (2009)
4. Donnelly, M.P., Nugent, C.D., Mason, S., McClean, S.I., Scotney, B.W., Passmore, A.P., Craig, D.: A Mobile Multimedia Technology to Aid Those with Alzheimer's Disease. IEEE Multimed. 17, 42–51 (2010)
5. Nugent, C., O'Neill, S., Donnelly, M., Parente, G., Beattie, M., McClean, S., Scotney, B., Mason, S., Craig, D.: Evaluation of Video Reminding Technology for Persons with Dementia. In: Abdulrazak, B., Giroux, S., Bouchard, B., Pigot, H., Mokhtari, M. (eds.) ICOST 2011. LNCS, vol. 6719, pp. 153–160. Springer, Heidelberg (2011)
6. O'Neill, S.A., Parente, G., Donnelly, M.P., Nugent, C.D., Beattie, M.P., McClean, S.I., Scotney, B.W., Mason, S.C., Craig, D.: Assessing task compliance following mobile phone-based video reminders. In: IEEE EMBC, pp. 5295–5298. IEEE Press, Boston (2011)
7. O'Neill, S.A., Mason, S.C., Parente, G., Donnelly, M.P., Nugent, C.D., McClean, S.I., Scotney, B., Craig, D.: Video Reminders as Cognitive Prosthetics for People with Dementia. Aging International 36, 267–282 (2011)
8. Zehna, P.W.: Probability Distributions & Statistics. Allyn & Bacon, Boston (1970)
9. Alzheimer's Society. Factsheet 436: The Mini Mental State Examination (MMSE). Alzheimer's Society, London (2012), http://www.alzheimers.org.uk/site/scripts/documents_info.php?documentID=121 (accessed October 02, 2012)

10. Rokach, L., Maimon, O.: Top-down induction of decision trees classifiers - a survey. IEEE Transactions on Systems, Man, and Cybernetics, Part C: Applications and Reviews 35(4), 476–487 (2005)
11. Quinlan, J.R.: Induction of Decision Trees. Mach. Learn. 1(1), 81–106 (1986)
12. Quinlan, J.R.: C4.5: Programs for Machine Learning. Morgan Kaufmann, San Mateo (1993)
13. Hall, M., Frank, E., Holmes, G., Pfahringer, B., Reutemann, P., Witten, I.H.: The WEKA Data Mining Software: An Update. SIGKDD Explorations 11(1), 10–18 (2009)
14. Witten, I.H., Frank, E., Hall, M.A.: Data mining: Practical Machine Learning Tools and Technologies, 3rd edn. Morgan Kaufmann, San Francisco (2011)
15. Hilbe, J.M.: Logistic Regression Models. Chapman & Hall/CRC Press, Boca Raton (2009)
16. O'Neill, S.A., McClean, S.I., Donnelly, M.P., Nugent, C.D., Galway, L., Young, T., Scotney, B.W., Mason, S.C., Craig, D.: Development of a Technology Adoption and Usage Prediction Tool for Assistive Technology for People with Dementia. Submitted to Interacting with Computers (under review)

The Discharge Planning Dilemma in the UK NHS: The Role of Knowledge Management

Nitya Kamalanathan[1], Alan Eardley[1], Caroline Chibelushi[1], and Paul Kingston[2]

[1] School of Computing, Staffordshire University, Stafford, UK
[2] Faculty of Health Sciences, Staffordshire University, Stafford, UK
{n.a.kamalanathan,w.a.eardley}@staffs.ac.uk

Abstract. This paper describes a research project that investigates and evaluates the role of Knowledge Management (KM) in discharge planning (DP) within the UK National Health Service (NHS). KM has been promoted in the NHS for a little more than 13 years. The paper shows that the popular press frequently reports problems associated with DP and examines some more reliable sources, concluding that more research into the phenomenon is needed. The factors that contribute to inadequate DP are summarised as conclusions to the paper. This is therefore an extract from a wider research project into the use of KM in DP, which is aimed at suggesting some causes, indicating some possible solutions and producing a KM framework to guide the DP and decision-making process. The paper indicates the current status of the research, which is continuing. It is hoped that this paper highlights the problem, summarises the research to date and stimulates further discussion of this important topic. Further publications will disseminate the developed solution.

Keywords: Knowledge management (KM), healthcare, discharge planning (DP), bed blocking, emergency readmissions, waiting lists, frameworks.

1 Introduction

The UK NHS has more than 1.2million employees working in 28 strategic health authorities (SHA), which manage 276 hospital trusts and 302 primary healthcare trusts (PHT). A typical healthcare scenario (e.g. an acute hospital or a care ward) can be considered as a typical example of an organisational system, with a collection of independent but interrelated elements or components organised in a meaningful way in order to accomplish an overall goal [29]. Just like any other system, a care ward is made up of subsystems having conventional components such as inputs, processes and outputs, all of which are components of a larger system (i.e. the hospital system) which is in turn a part of larger healthcare system. These systems and subsystems are clearly interdependent and inter-related. It is therefore important to understand healthcare subsystems in order to gain a deeper insight into the functioning of the system [5]. The research project that this papers desribes therefore focuses on analysing the hospital system in terms of its structure and process in terms of:

G. Huang et al. (Eds.): HIS 2013, LNCS 7798, pp. 172–185, 2013.

- The components themselves (e.g. patients, nurses) and their roles in the system;
- The relationship between the components and their interaction (e.g. nurses care for patients);
- The boundaries of the system or its extent and scope (e.g. where an admission ward hands over to an operating theatre) or where patients are discharged;
- How the system deals with and adapts to changes within the organisation (e.g. emergency admission or an outbreak of an infection).

By having a sound understanding of the structure and functioning of the system, the practitioner is able to discover new knowledge to update existing knowledge and use that knowledge in decision-making processes such as patient discharge planning (DP). Post-treatment care is obviously essential to a patient's complete healthcare pathway and DP plays a key factor in a patient's convalescence. The factors that play instrumental roles in DP are many and varied but usually involve patients being transferred from one care environment to another (e.g. from an admission ward to an acute ward). Careful planning and a clear decision-making framework are vital to the smooth flow of patients from admission to discharge at the end of the treatment period. Patient discharge can be considered to be the beginning of convalescence. The 'system view' [19] suggests that as such DP is a key part of the overall process and is not an isolated or final event. This view would extend the 'system boundary' to include what happens to a patient after discharge, to prevent unwanted readmissions. This implies the involvement in the system of patients, their families and carers involved in their post-treatment and recovery period. As such it has implications for the provision of resources in the healthcare, social care and other support service sectors and warrants this research to improve its efficiency and effectiveness.

A smooth DP process facilitates patients moving from one healthcare setting to another, or going home. It begins on admission and is a multidisciplinary process involving physicians, nurses, social workers, and possibly other health professionals [4]. The aim of DP is therefore to enhance the continuity of care and can have significant implications for a patient's wellbeing and recovery, the efficient use of medical resources and streamlined interconnecting processes within the hospital setting. The complexity of the discharge process implies that careful planning is needed to make it more effective [36]. Recent years have witnessed significant advances in medical informatics to increase productivity and efficiency in healthcare [11]. Some parts of the NHS are currently faced with the problem of 'islands of information'[1] related to the existence of organisational 'silos'[2]. In some cases it is suggested that very little knowledge is shared between these silos. This leads to the foundation of this paper, which is to examine the role of KM in an integrated 'cross-silo' approach to using shared knowledge to create appropriate patient discharge pathways. KM therefore forms a bridge between these 'islands of information' [28].

---

[1] IT applications that were originally developed to solve localised problems, but which do not communicate with other applications in the same IT infrastructure.

[2] Parts of the organisation (e.g. departments, functions) that are separate in terms of processes, communication and policies.

2   Knowledge Management

Knowledge is a multifaceted concept with multi-layered meanings [26] [27]. Due to this nature, it has become important to manage knowledge in order to drive performance by ensuring that the relevant knowledge is delivered 'to the relevant person in the right place in a timely fashion' [13]. Apart from existing in the human mind (i.e. tacit knowledge), knowledge can exist in physical records (i.e. explicit knowledge), such as patient records and medical notes, which can be shared and accessed more readily whether in paper or computerised form. The major focus of KM in healthcare is on creating environments for 'knowledge workers' to create, leverage and share knowledge and for this to happen effectively KM requires deep-rooted behavioural and strategic change. From this point of view KM represents an evolution of the move towards greater personal and intellectual freedom [8] empowering individuals to engage more actively in their work by sharing ideas, thoughts and experiences [11]. Once the knowledge has been discovered, storing it, reusing it and generating new knowledge from it is important to 'adding value' to data to create shared knowledge.

Continued progress in technology development makes sharing knowledge easier, and the Internet and collective portals makes knowledge accessible to a wider range of people [8]. The rise of networked computers has made it easier and cheaper to codify, store and share knowledge [15]. There is no shortage of technologies to aid in managing knowledge in a healthcare environment [10], rather the prevalence of such technologies can create confusion. The goal of KM is to enhance the performance of a process (e.g. discharge planning) by providing efficient access to knowledge, experts and communities of practice. It aims to prioritise, share, consolidate and provide consistent and accurate information and performance indicators in order to help with efficient decision making processes. As workers in a 'knowledge intensive environment', healthcare professionals inevitably hold a considerable amount of experiential knowledge, which may be used to solve day to day problems (e.g. decisions on patient discharge). It is important that the knowledge used to solve such problems is captured, shared and reused in order to prevent the lack of 'nourishment' (i.e. update and replenishment) of that knowledge [21] and to improve 'knowledge of context'. The 'knowledge process' in a healthcare environment can be used to increase collaboration with clinicians, nursing staff and social service agencies and for purposes of innovation or process improvement. Updating knowledge assets cultivates the collective knowledge in a healthcare environment, enriching effective management, smoothing the flow of knowledge and enabling better problem solving [21] [22] and increasing 'knowledge potential'.

KM when applied effectively may result in increased efficiency, responsiveness, competency and innovation [12] which is a source of superior performance in potentially critical applications [31] such as patient discharge. The challenge is therefore to create a KM system that can 'acquire, conserve, organise, retrieve, display and distribute what is known today in a manner that informs and educates, facilitates the discovery of new knowledge and contributes [20] to the benefit of the organisation. KM can therefore be looked at as an integrating practice that offers a framework for balancing the technologies and approaches that provide value in making decisions and carrying out actions [37]. It ties them together into a seamless whole by aligning organisational information and practices with the organisation's objectives, fits into employee's daily work activities, manages content effectively, and considers the potential opportunities associated with sharing knowledge with external agents [12].

3   Knowledge Management in Healthcare

The healthcare industry has been called 'data rich while knowledge poor' [22] as its functions hold large amounts of data (e.g. patient records, outcomes of surgery and medical procedures, clinical trial data etc.) and yet the knowledge potential of many actions is yet to be fully exploited as much of the data is not being translated into knowledge (i.e. there is low added value) to provide a wider context, a deeper understanding and to help with strategic decision making [9]. Knowledge appears to be underutilised at the point of care and at the point of need [9] inhibiting the ability of personnel with the relevant experience to 'harvest' knowledge and provide a clearer understanding of the process and the factors involved by providing 'a window on the internal dynamics of the healthcare enterprise' [36]. Multidisciplinary teams working in healthcare harvest the personal expertise that is essential to patient safety, learn from it, adapt it to local situations and individual patients, and distribute it via reliable networks to the people caring for the patients so that they can use it to improve the quality of care delivered [25]. The knowledge that a typical healthcare application possesses is a 'high value form of information' that allows sharing of the lessons learned from past experiences [36] (e.g. knowing what factors to take into account when planning the discharge of a patient) and improves the knowledge potential of the process in future by improving knowledge of the knowledge context.

A fundamental challenge faced by clinical practitioners and healthcare institutions is the ability to interpret clinical information and to make potentially lifesaving decisions while dealing with large amounts of data [10]. Clinical practise is not only quantitative, but also very much qualitative. The tacit knowledge acquired by clinicians and nurses over the years and mainly through experience represents a valuable form of clinical knowledge [24]. KM in Healthcare involves understanding diseases, hospital systems and most importantly patients [24]. Levenstein et al. argue that clinical methods exist for understanding diseases and illnesses but clinical methods or models are not so readily available for understanding patients. When quantitative and qualitative methods complement each other, and when various modalities of knowledge are used, a holistic view of a situation is best obtained thus leading to efficient decision making [20]. KM strategies can be broadly classified into codification (where knowledge is identified, captured, indexed and made available) and personalisation, where tacit knowledge is shared by means of discussion, effective communication and a multidisciplinary approach, allowing for creative problem solving [20]. In Healthcare, the use of both strategies according to the different scenarios is felt to be advisable. When dealing with routine cases, the codification strategy can be applied and when dealing with a situation where a more creative solution is required, the personalisation strategy can be applied [20]. This approach, however, usually only works when the required knowledge is shared (i.e. processed) successfully. For example the National Institute for Clinical Excellence (NICE), in framing its guidelines, has noted a lack of willingness to share knowledge on the part of doctors who could potentially contribute to the guidance it gives [25].

4   Knowledge Management for Discharge Planning

A hospital is a dynamic environment, with changes taking place rapidly as patients move from one ward to another and treatments are carried out over time. Similarly, DP involves changes from a stable temporal state to another with an element of unpredictability of what is going to happen next [21]. In this context the past experiential knowledge of doctors and nurses is useful in assessing situations, deciding on plans and making critical decisions as their knowledge can be reconfigured and extended to fit the new situation and provide a personalised approach to assessing a patient and his or her journey along codified guidelines [21]. KM may have the potential to remove the bottlenecks, to improve the DP process, mapping and identifying possible opportunities for improvement [29]. Understanding what knowledge is relevant to a given situational decision is crucial to this process and a decision can never be completely separated from the context in which it is made [12]. This implies that in a hospital setting when looking at DP the interrelated factors need to be considered in the context of the knowledge process [23]. Discharge takes place when an in-patient3 leaves an acute hospital and returns home, is transferred to a rehabilitation facility or an after-care centre such as a nursing home [25].

DP should commence as early as possible in order to facilitate a smooth discharge process. Discharge guidelines have been prescribed by the Department of Health (DoH) and the different trusts create their discharge plans in the form of a discharge flow chart or process map following these guidelines. Several attempts have been made at improving DP and reasonable improvements have been identified. Several of the methods that have been identified in the primary research in two UK hospital trusts include the following:

- DP commences on admission
- Patient and carer involved in the decision making process
- A clinical management plan where an expected date of discharge is predicted based on actual performance in the ward or, on benchmarking information from past cases;
- Multidisciplinary teams make a decision based on experience during their meetings;
- A bed management system which stores information on beds occupied, and a weekly meeting that decides the discharge date for patients.

---

3 Outpatients are usually treated 'on the spot' and medication is prescribed on the same day, so the DP process does not usually apply (NHS, 2012).

All of these methods involve KM. A rough DP is currently drafted for patients upon entry to hospital according to their diagnosis, and a tentative discharge date provided in line with recommendations. Changes are made over the course of the patient's stay and records are manually updated by nurses, upon instruction by the doctors. This sometimes results in confusion and even disagreement on discharge dates by different doctors (e.g. when treating the patient for different symptoms) and nurses (e.g. when a change of shift occurs). Patient discharge planning requires looking at the system as a whole and not as isolated units. The discharge plan indicates patient and carer involvement, however very little indication has been provided of the nature of involvement. Clear guidelines are not present, as to what information needs to be collected about a patient, what information needs to be stored and reused, and the lessons learnt.

5   The Current DP Dilemma in the NHS

The UK NHS is facing the problem of managing patient discharge while having to meet waiting time, treatment time and bed usage targets [10]. Patient discharge is currently being driven by quantitative measures such as targets (e.g. to reduce 'bed-blocking'[4]) and the problem resulting from this situation has received a great deal of attention from the popular press recently and political capital has been made from this [10]. Targets are often given priority while a patient's quality of after-care is compromised [32]. The implication of being target-driven (rather than knowledge driven) is that the healthcare system fails to consider the factors that affect the effective recovery of a patient *after* treatment and discharge [6]. Hospitals focus on accomplishing and achieving internal targets, resulting in compromising patient safety and well-being after discharge. The exact situation with regard to patient discharge and readmissions is not really well established, as there are variations in discharge methods between trusts. However, it is reported in the popular press that doctors have to make quick decisions about patients just to 'get the clock to stop ticking' [32] [33] [39] resulting in deteriorating trust between doctors and patients. More reliably, doctors find themselves torn between meeting targets or providing their sick patients with the best treatment. These claims in the assorted news media have been reaffirmed by Andrew Lansley the Secretary of State for Health in the UK Government who in a speech in December 2011 stated that [7]:

> '*The NHS is full of processes and targets, of performance-management and tariffs, originally, all designed to deliver better patient care, but somewhere along the line, they gained a momentum of their own, increasingly divorced from the patients who should have been at their centre*'.
>
> (Guardian 7 December 2012)

---

[4] Bed blocking occurs when patients are kept in hospital wards after they should be discharged, pending a decision to send them home or into care. It is considered to be a major problem in some parts of the HNS, particularly with elderly patients. The opposite may involve discharging patients before they are ready to meet a target. See http://www.communitycare.co.uk/Articles/09/02/2011/116235/portsmouth-makes-the-case-for-joint-working-to-cut-bed-blocking.htm

Several factors result in the current inadequate DP. These factors are internal and external to the NHS along with psychosocial factors of the patient and family. The reason for understanding the factors behind inadequate DP is to be able to analyse the factors causing the problem and diagnose the problem systematically. A comparison can then be made between the factors along with the results obtained from the primary research, followed by a catalogue of possible solutions underpinned by KM. This will then lead to making a diagnosis i.e. the proposed KM model. A root cause analysis [17] was carried out to analyse the factors contributing to inadequate DP and diagrammatically represented in the form of a 'fishbone diagram' as seen in Figure 2. The results obtained from primary and secondary research were used in the analysis. As seen in the fishbone diagram, the discharge of a patient is a complex process, with various inter-related factors. A carefully designed discharge plan underpinning the theory of KM can ensure that hospital resources are utilised more efficiently, it will encourage better inter-department communication and ensure that tacit knowledge is made explicit to make more informed decisions about patient discharge. It is believed that this in turn will allow for better coordination of the external factors involved and will give hospital personnel more time to inform patients and their families about their situation, thus addressing the psychosocial factors. It is indicated that many adverse events occur during discharge. Many of these events are preventable and are therefore viewed as errors, while others are undetectable adverse events. Therefore with more sharing of knowledge between personnel in the hospital, an understanding of the knowledge needed to make a decision and understanding, utilising and capturing the knowledge gained from a certain procedure or step in the admission and discharge pathway can lead to improved DP, which can be personalised to a patient. Patient participation in the discharge process is believed to help reduce potential readmissions and delayed discharge. Patient participation in the discharge process is a legally stated right in the United Kingdom and therefore more active participation of patients is encouraged [38]. The failure to assess a patient's care needs correctly can result in a disproportionate delay in patients being discharged [16].

The causes of inadequate DP have been examined, and these causes are shown to have a number of negative effects. Several consequences of inadequate DP faced by the NHS include; delayed discharge and increased emergency readmissions. These result in bed blocking and long waiting lists. Overall due to these reasons admissions, transfer and discharge processes are negatively affected as a consequence. The problems caused by inadequate DP have been identified in the secondary and primary research and are summarised succinctly in Figure 3. The number of patients readmitted to hospitals through Accident and Emergency (A&E) departments within 28 days of being discharged has risen steadily from 359,719 in 1998 to 546,354 in 2008 [40]. While in 2010 more than 660,000 patients were re-admitted to hospital within 28 days of discharge [30].

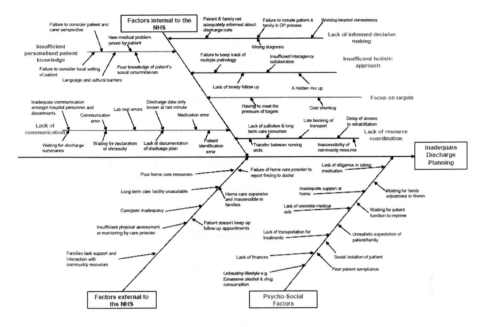

Fig. 1. Root Cause Analysis of factors resulting in inadequate DP
Source: [35, 38]

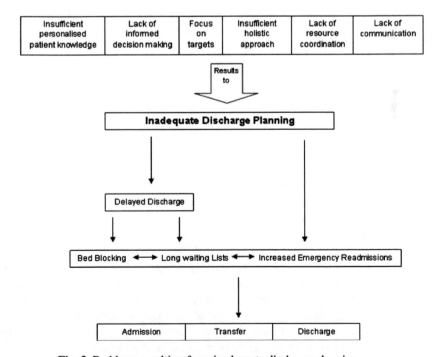

Fig. 2. Problems resulting from inadequate discharge planning
Source: [35]

According to statistics provided by the Department of Health, in England in 2010-2011 the total number of patients who were readmitted was 561,291. According to the statistics, readmission rates in England have been rising since 2001 -2002 to 2010 – 2011. Figure 4 follows the increasing trend of the percentage of patients readmitted for treatment to UK acute hospitals within 30 days of discharge and a 'line of best fit' shows the regularity (and therefore the predictability) of the rise.

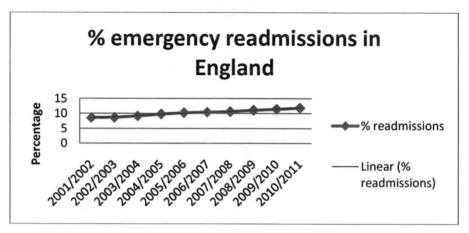

Fig. 3. Emergency 3 Readmissions in England as a Percentage of Admissions

The problem of inadequate DP does not just concern readmissions, however. 'Bed-blocking' due to delayed discharge has equivalent negative implications. It is reported by the NHS confederation that one in four patients are occupying beds when they could be recovering at home [16], which results in longer waiting lists, loss of confidence in the NHS and escalating expenditure. The average number of patients and days of delayed discharge per month in England for the year 2012 according to the Department of Health was 3997 patients and 114,386 days respectively.

Approximately £250m was spent on 'delayed discharges' between August 2010 and the end of 2011, amounting to £550,000 a day [40]. Apart from the financial implications the delay in discharge is clearly not good for the well-being of patients and the morale of their relatives and wastes valuable hospital resources. The King's Fund reports that if it was better organised the NHS could reduce the number of overnight stays by 2.3 million, freeing up 7,000 beds and saving the NHS nearly £500m a year. [30] Mike Farrar, the Chief Executive of the NHS Confederation indicates that these problems are the result of an 'outdated hospital model of care' [3] while a breakdown in communication may also be a possible contributory cause [40]. Many older patients face the brunt of delayed discharge [13] as due to a lack of communication between the NHS and social care homes, they are forced to stay in hospital, causing longer waiting lists for other patients who are seeking urgent treatment [42]. The reasons for the dilemma as described in the previous section are clearly a result of inadequate support for DP among NHS staff, including physicians, nurses, social workers, and possibly other health professionals [46]. The aim of DP is to enhance the continuity of care while optimising use of healthcare resources. DP also has significant implications for a patient's recovery, the effectiveness of hospital management processes and the efficient use of medical resources. The complexity of the discharge process implies that careful planning is needed to make the process effective [5] and it is recognised that the problem may currently lie in a lack of appropriate DP upon admission. A lack of sufficient knowledge of the patient's circumstances [41] upon admission results in one of three problems i.e. problems with admission, problems with transfer and problems with discharge as previously indicated in Figure 3. Patient discharge is fragmented among various caregivers [4].

DP commences on admission, the patient and carer are involved in the DP process, their medication, nutrition, and aftercare is arranged prior to discharge. These are the steps that are taking place on paper in NHS Trusts. However, the increased emergency readmissions, delayed discharges, bed blocking and long waiting lists that increasingly occupies the media's attention indicates that the guidelines and process may not be optimal. The primary research in two UK NHS Trusts that informs this paper suggests that the approach that is currently being followed could be improved by being knowledge-based rather than process-focused. The research therefore investigate the statistics as reported in the media, the statistics published by the DoH and the analyses the results obtained from Primary research in two UK NHS Trusts involved in the primary research. The results obtained from the three sets are analysed, a comparison is made and possible themes are identified to indicate the current knowledge gaps that exist in the way DP is currently carried out. This is basically a Grounded Theory approach and will produce a KM framework that is soundly based on KM theory. The KM model will ensure that the people, processes and technologies are aligned in such a way that the right people get the right information at the time when it is needed and will break down the current silos that exist between departments of the same trust and between separate NHS Trusts.

Despite having guidelines, a clear delineation of discharge responsibilities often does not exist or is not communicated properly resulting in repetition of procedures or gap [4]. An example of this would be the lack of coordinating with the pharmacy or the physiotherapy department upon discharge of a patient which can result in a patient's discharge being delayed, or a patient going home without medication and having to return a few days later to collect the medication, or a patient having to wait a few days later for their physiotherapy schedule [14].

6   Conclusions

The research in two UK hospital trusts shows that inadequate knowledge of a patient's individual circumstances (e.g. home care, state of mind) are indentified as major contributors to a significant number of unwanted readmissions. This in turn leads to decisions being made without the necessary information being available for inclusion in the decision-making process that informs DP. In other words, the patient records (e.g. electronic patient records or EPR) do not contain all of the qualitative information to act as valid knowledge assets (e.g. a discharge plan) or resources to support the knowledge action of DP (i.e. making the decision to discharge) [18]. In addition, it is found that a reliance on quantitative measures (e.g. meeting discharge targets) may be contributing to patients being discharged before they have been observed sufficiently to ensure their satisfactory transfer to an aftercare environment. The current DP process in the study appears to be insufficiently 'holistic' (i.e. it fails to take all the necessary factors into consideration) and indicates that resources (including knowledge resources) were not co-ordinated closely enough to bring all the necessary factors to bear on the problems of inadequate DP. Most of all, it is noted that communication across the 'silos' in a typical hospital environment were inhibiting the full use of knowledge in the DP process.

The results of inadequate DP are noted in the research (see Figure 2) and delayed discharge (as reported in the popular press) is found to be the most visible problem. Less visible, but problematical none the less, are bed-blocking, resulting in unnecessarily long waiting lists and a high level of unwanted readmissions within 28 days of discharge, often complicating the issue by involving overcrowded A&E departments (i.e. 'emergency readmissions'). The proposed model intends to apply a more holistic approach; it intends to peruse quantitative and qualitative data in order to make more informed decisions. Furthermore it intends to encourage the sharing of information between the current silos, allowing for more informed decisions to be made. The processes or departments that are adversely affected are admissions to hospital, transfer between hospital departments or wards and discharge. The preliminary findings from the research indicate, subject to validation, that the judicious use of KM can improve the DP process and ameliorate the current negative effects in all of these areas.

References

1. Abidi, S.S.R.: Healthcare Knowledge Management: The Art of the Possible, pp. 1–20 (2008)
2. Abidi, S.S.R.: Knowledge management in healthcare: towards "knowledge-driven" decision-support services. International Journal of Medical Informatics 63, 5–18 (2001), http://ac.els-cdn.com/S1386505601001678/ 1-s2.0-S1386505601001678-main.pdf?_tid=553fb9d2-25b7-11e2-bf81-00000aacb362&acdnat=1351948191_25c6656dcc2be7462b339ac9fcf9957c (accessed January 28, 2013)
3. Adams, S.: 'Intolerable' bed-blocking crisis threatens NHS. The Telegraph (February 2011), http://www.telegraph.co.uk/health/healthnews/8343879/ Intolerable-bed-blocking-crisis-threatens-NHS.html (accessed January 28, 2013)
4. Anthony, D., Chetty, V.K., Kartha, A., McKenna, K., DePaoli, M.R., Jack, B.: Re-engineering the Hospital Discharge: An Example of a Multifaceted Process Evaluation. In: Advances in Patient Safety: From Research to Implementation, vol. 1-4. AHRQ Publication Nos. 050021 (1-4). Agency for Healthcare Research and Quality, Rockville (February 2005), http://www.ahrq.gov/qual/advances/ (accessed January 28, 2013)

5. Ashmos, P., Huber, P.: The Systems Paradigm in Organization Theory: Correcting the Record and Suggesting the Future. JSTOR: The Academy of Management Review 12(4), 607–621 (1987), http://jaylee.business.ku.edu/MGMT916/PDF/AshmosHuber1987AMR.pdf (accessed January 28, 2013)

6. Bali, R.K., Dwivedi, A., James, A.E., Naguib, R.N.G., Johnston, D.: Towards a holistic knowledge management framework for healthcare institutions. In: Proceedings of the Second Joint EMBSBMES Conference, pp. 1894–1895. IEEE, Houston (2002), http://ieeexplore.ieee.org/stamp/stamp.jsp?arnumber=01053081 (accessed January 28, 2013)

7. Boseley, S.: Andrew Lansley launches 60 NHS "patient outcome measures" measures. The Guardian (Decmber 2011), http://www.guardian.co.uk/society/2011/dec/07/andrew-lansley-patient-outcome-measures/print (accessed January 28, 2013)

8. Chunsheng, T.: The Research on Knowledge Management and Knowledge Portal Framework based on Agent Mechanism, pp. 263–267 (2000)

9. Copper, A.: We Haven"t Got a Plan, so What Can Go Wrong? Where is the NHS Coming from? In: Bali, R.K., Dwivedi, A. (eds.) Healthcare Knowledge Management, pp. 221–232. Springer Science+Business Media (2007)

10. Dwivedi, A., Bali, R.K., James, A.E., Naguib, R.N.G., Johnston, D.: Merger of knowledge management and information technology in healthcare: opportunities and challenges. In: IEEE CCECE 2002. Canadian Conference on Electrical and Computer Engineering. Conference Proceedings (Cat. No.02CH37373), pp. 1194–1199. IEEE (2002), http://ieeexplore.ieee.org/lpdocs/epic03/wrapper.htm?arnumber=1013118 (accessed January 28, 2013)

11. Eardley, W.A., Czerwinski, A.: Knowledge Management for Primary Healthcare Services. In: Bali, R., Dwivedi, A. (eds.) Knowledge Management: Issues, Advances and Successes. Springer Science + Business Media (2007)

12. Fontaine, M., Lesser, E.: Challenges in managing organizational knowledge (2002), http://www-935.ibm.com/services/us/imc/pdf/g510-3234-00-esr-managing-organizational-knowledge.pdf (accessed January 28, 2013)

13. Foss, I.C., Hofoss, D.: Elderly persons' experiences of participation in the hospital discharge process. Patient Education and Counseling 85(1), 68–73 (2011)

14. Godden, S., McCoy, D., Pollock, A.: Policy on the rebound: trends and causes of delayed discharges in the NHS. Journal of the Royal Society of Medicine 102(1), 22–28 (2009), http://www.ncbi.nlm.nih.gov/pubmed/19147853

15. Hansen, T.M., Nohria, N., Tierney, T.: What's Your Strategy for Managing Knowledge? Harvard Business Review, 1–11 (1999), http://consulting-ideas.com/wp-content/uploads/Whats-your-strat-art.pdf (accessed January 28, 2013)

16. Improving Patient Flow at Bon Secours Venice Hospital. Institute for Healthcare Improvement (2011), http://www.ihi.org/knowledge/Pages/ImprovementStories/ImprovingPatientFlowBonSecoursVeniceHospital.aspx (accessed January 28, 2013)

17. Ishikawa, K. (Trans: J. H. Loftus); Introduction to Quality Control. 3A Corporation, Tokyo (1990)

18. Knott, L.: Records, Computers and Electronic Health Record (2012), http://www.patient.co.uk/doctor/Records-Computers-and-Electronic-Health-Record.htm (accessed January 7, 2013)

19. Laszlo, E.: The Systems View of the World. Hampton Press, N.J (1996)

20. Levenstein, H.J., McCracken, C.E., Ian, M.R., Stewart, A.M., Brown, B.J.: The Patient-Centred Clinical Method. Family Practice 3(1), 24–30 (1986), http://fampra.oxfordjournals.org/content/3/1/24.full.pdf+html (accessed January 28, 2013)

21. Liao, S.: Problem solving and knowledge inertia. Expert Systems with Applications 22(1), 21–31 (2002), http://linkinghub.elsevier.com/retrieve/pii/S095741740100046X (accessed January 28, 2013)

22. Mills, A.M., Smith, T.A.: Knowledge management and organizational performance: a decomposed view. Journal of Knowledge Management 15(1), 156–171 (2011), http://www.emeraldinsight.com/10.1108/13673271111108756 (accessed October 10, 2012)

23. Mishler, E.G.: Meaning in Context: Is There Any Other Kind? Harvard Educational Review, 1–19 (2012)

24. NHS: About the NHS. NHS (2012a), http://www.nhs.uk/NHSEngland/thenhs/about/Pages/overview.aspx (accessed January 28, 2013)

25. NHS: Leaving Hospital. NHS (2012b), http://www.nhs.uk/nhsengland/aboutnhsservices/nhshospitals/pages/leaving-hospital.aspx (accessed January 28, 2013)

26. Nonaka, I., Lewin, A.Y.: Dynamic Theory Knowledge of Organizational Creation. Organization Science 5(1), 14–37 (1994), http://www.e-cam-pus.uvsq.fr/claroline/backends/download.php?url=L05vbmFrYTk0LnBkZg%3D%3D&cidReset=true&cidReq=MDSDM105 (accessed January 28, 2013)

27. Nonaka, I., Toyama, R.: The knowledge-creating theory revisited: knowledge creation as a synthesizing process. Knowledge Management Research & Practice 1(1), 2–10 (2003), http://www.palgrave-journals.com/doifinder/10.1057/palgrave.kmrp.8500001 (accessed November 2, 2012)

28. Polanyi, M.: Tacit Knowing. In: Polanyi, M. (ed.) The Tacit Dimension, pp. 3–25. Doubleday & Company, New York (1966)

29. Pratt, J., Gordon, P., Lamping, D.: System that knows itself. In: Working Whole Systems: Putting Theory Into Practice in Organisations, pp. 22–24. Radcliffe Publishing (1999)

30. Ramesh, R.: One in four hospital patients "could be recovering at home". The Guardian (Decmber 2011), http://www.guardian.co.uk/society/2011/dec/29/hospital-patients-discharge-bed-blocking (accessed January 28, 2013)

31. Reychav, I., Weisberg, J.: Bridging intention and behavior of knowledge sharing. Journal of Knowledge Management 14(2), 285–300 (2010), http://www.emeraldinsight.com/10.1108/13673271011032418 (accessed October 10, 2012)

32. Roberts, M.: Hospitals to face financial penalties for readmissions. BBC News (June 8, 2010), http://www.bbc.co.uk/news/10262344 (accessed January 28, 2013)

33. Ross, T.: Rise in "bed blocking" is costing the NHS £500,000 every day. The Telegraph (December 2011), http://www.telegraph.co.uk/health/elderhealth/8929747/Rise-in-bed-blocking-is-costing-the-NHS-500000-every-day.html

34. Roy, R., Rey, F.M., Wegen, B.V., Steele, A.: A Framework To Create Performance Indicators In Knowledge Management. In: Proc. of the Third Int. Conf. on Practical Aspects of Knowledge Management (PAKM 2000), Basel, Switzerland, pp. 30–31 (2000), http://sunsite.informatik.rwth-aachen.de/Publications/CEUR-WS/Vol-34/ (accessed January 28, 2013)

35. Sg2 Healthcare Intelligence. Reducing 30-Day Emergency Readmissions (2011), http://www.hsj.co.uk/Journals/2/Files/2011/6/15/Sg2_ServiceKit_Reducing30-DayReadmissions.pdf (accessed January 28, 2013)
36. Shepperd, S., McClaran, J., Phillips, C., Lannin, N., Clemson, L., McCluskey, A., Cameron, I., Barras, S.: Discharge planning from hospital to home. The Cochrane Library (1), 1–75 (2010), http://onlinelibrary.wiley.com/doi/10.1002/14651858.CD000313.pub3/pdf (accessed January 28, 2013)
37. Sharp, P.J., Eardley, W.A., Shah, H.: Visual Tools within MakE - A Knowledge Management Method. Electronic Journal of Knowledge Management 1(2), 177–186 (2003)
38. Social Care Institute for Excellence. SCIE Research Briefing 12: Involving individual older patients and their carers in the discharge process from acute to community care: implications for intermediate care, http://www.scie.org.uk/publications/briefings/briefing12 (accessed January 28, 2013)
39. The Press Association, Bed blocking 'on the increase.' Nursing Times. net (2011), http://www.nursingtimes.net/nursing-practice/clinical-zones/critical-care/bed-blocking-on-the-increase/5023593.article (accessed January 28, 2013)
40. Triggle, N.: Organise urgent care better, NHS told. BBC News (August 2012), http://www.bbc.co.uk/news/health-19181905 (accessed November 8, 2013)
41. Wickramasinghe, N., Mills, G.L.: Knowledge Management Systems: a Healthcare Initiative with Lessons for us all. Knowledge Management, 763–774 (2001)
42. Winnett, R.: Scandal of NHS "production line" as readmissions soar. The Telegraph (December 29, 2011), http://www.telegraph.co.uk/health/healthnews/8983505/Scandal-of-NHS-production-line-as-readmissions-soar.html (accessed January 28, 2013)

Using Semantic-Based Association Rule Mining for Improving Clinical Text Retrieval

Atanaz Babashzadeh, Mariam Daoud, and Jimmy Huang

Information Retrieval and Knowledge Managment Research Lab
School of Information Technology
York University, Toronto, Canada
{atanaz,daoud,jhuang}@yorku.ca

Abstract. Association rule (AR) mining has been widely used on the electronic medical records (EMR) for discovering hidden knowledge and medical patterns and also for improving the information retrieval performance via query expansion. A major obstacle in association rule mining is that often a huge number of rules are generated even with very reasonable support and confidence. The main challenge of using AR in information retrieval (IR) is to select the rules that are related to the query, since many of them are trivial, redundant or semantically wrong. In this paper, we propose a novel approach to modeling medical query contexts based on mining semantic-based AR for improving clinical text retrieval. We semantically index the EMR with concepts of UMLS ontology. First, the concepts in the query context are derived from the rules that cover the query and then weighted according to their semantic relatedness to the query concepts. The query context is then exploited to re-rank patients records for improving clinical retrieval performance. We evaluate our approach on the medical TREC dataset. Results show that our proposed approach allows performing better retrieval performance than the probabilistic BM25 model.

1   Introduction

Due to increasing volume of digitalized medical patient records, the need for advanced information retrieval systems increases. Our work lies in the framework of the 2011 TREC medical record challenge. The goal of this track is to foster research on content-based retrieval from the free-text fields of electronic medical records.

Many participants in TREC Medical Record track proposed different techniques to improve their system's retrieval task. In [5] author's participation relies on query expansion technique using Rocchio's algorithm coupled with gender and age filtering and semantic query expansion using disease synonyms. This approach did not have a significant impact on retrieval performance. The lack of significant improvement on the overall retrieval results shows that their method has limited semantic capabilities. In [6] authors' participation relies on applying part-of-speech tagging and UMLS concept extraction at the sentence level using bi-directional greedy dictionary matching for noun phrases and query expansion by inclusion of all concepts appearing below the concept of interest in the UMLS hierarchy, capped at the

G. Huang et al. (Eds.): HIS 2013, LNCS 7798, pp. 186–197, 2013.

maximum of 100 concepts. Performance measures suggest that their approach was not promising, possibly due to appending too specific concepts to the original query. In [3] authors participation relies on query expansion via multiple lexicons and domain knowledge rules and performance measures imply limited impact on retrieval performance, possibly due to limited semantic capabilities of expansion terms.

In this paper, we focus on the use of AR mining for improving clinical text retrieval. In the biomedical field, association rule mining presents a promising technique for finding hidden patterns in a medical dataset and for improving the information retrieval performance via query expansion. It was first introduced in 1993 by Agrawal et al. [1]. Its objective is to discover all co-occurrence relationships, called associations, among data items under the form of implications if X then Y, denoted as $X \rightarrow Y$, where X and Y are named the antecedent and the consequent of the rule respectively. Association rule mining has been widely applied on a medical data for diagnosis or symptom prediction [19,16], gene expression and cell type prediction [10] classification purposes [14] and information retrieval [8,11,13,24, 25,26]. The usefulness of association rule technique is strongly limited by the huge amount and the low quality of delivered rules. Current data mining techniques can efficiently generate association rules that are statistically significant to the source dataset where rules satisfy the user-specified minimum support and minimum confidence. The use of AR in IR is challenging since many rules are trivial, redundant, semantically wrong and conflict with common sense or basic domain knowledge, or already known by end-users.

The main challenging issue in using AR in IR is to select the best rules with respect to the query. In most of previous work, the rules that contain the query terms in the antecedents are selected. Most of previous work lacks of semantics for selecting the relevant rules, which is more necessary in medical domain. In this paper, we propose the use of semantic relatedness measure using the UMLS semantic network for selecting concepts from rules that semantically cover the query. The selected rules are used to form a medical query context, which is used for re-ranking the search results.

This paper presents a novel medical query context modeling based on association rule mining to improve clinical text retrieval performance. The key unique contributions in our paper concern (1) a semantic-based AR mining for modeling a medical query context and (2) a semantic-based concept weighting schema to weight concepts semantically in relation to the query. Query context modeling relies on deriving the rules that cover the query and then weighted according to their semantic relatedness to the central concept in the query. The query context is then exploited in the patient records re-ranking for improving clinical retrieval performance.

This paper is organized as follows: Section 2 explains association rule mining used in this study, Section 3 describes semantic query context modeling for clinical text retrieval, Section 4 explains the experimental settings, Section 5 describes our experimental results and Section 6 concludes the paper.

2   Association Rule Mining

One of the classic association rule mining algorithms is the Apriori algorithm. The Apriori algorithm is an algorithm for mining frequent itemsets, it attempts to find frequent itemsets that have minimum support where any subset of a frequent itemset must also be a frequent itemset [1]. Given a transaction database and support threshold, standard association rule algorithm has two phases; first phase finds all itemsets satisfying minimum support, it starts with generating frequent one-itemsets and proceeds to two-itemsets and so on, until there are no more frequent itemets. Second phase generates all rules with support and confidence above specified threshold.

Based on the choice of the minimum support and minimum confidence, Apriori algorithm can have the disadvantage of being very slow and producing huge or insignificant amount of information therefore most of the discovered rules are not useful since they may contain redundant information, irrelevant information or they may illustrate insignificant knowledge. In general, the main issues with discovering association rules in medical data are: irrelevancy of most of the discovered association rules, appearance of relevant association rules in very low support and discovery of enormous amount of rules at low minimum support. To address this issue TopKRules algorithm has been introduced. TopKRules algorithm [7] mines the top-k association rules; k is specified by users and represents number of association rules to be found. Experimental results show that TopKRules algorithm has excellent performance and scalability where the number of generated rules can be controlled. Another advantage of this algorithm is that it mines the top-k rules that meet specified confidence, since minsup is more difficult to assign due to its dependency on database characteristics and nature.

The inputs to TopKRules algorithm are a transaction database, minconf and a number K specified by the user, which corresponds to the number of rules to be returned. The algorithm starts with assigning 0 to internal minsup variable and searching for rules. Once a rule is found it is added to a list of rules sorted by their support. When K valid rules are found, the internal minsup is raised to lowest support of association rules in the list. New generated rules based on new minsup are added to the list and rules not satisfying the new minsup are removed from the list. This process continues until no more rules are found. The algorithm relies on rule expansion approach, it starts with generating rules with single antecedent item and single consequent item, and then it scans the transaction database in order to find single items that can be appended to left or right side of the association rule

3   A Semantic Query Context Modeling for Clinical Text Retrieval

Our approach consists of (1) modeling a semantic query context and (2) exploiting it in the re-ranking process of initial clinical search results. A prerequisite step in our work is the semantic representation of queries and patients records. This section presents an overview of the UMLS Metathesaurus used for semantic indexing, the query context modeling and the re-ranking process.

3.1   The UMLS Metathesaurus

The UMLS (Unified Medical Language System) is a set of health and biomedical dictionaries, standards and software tools that can facilitate biomedical and health related developing applications such as electronic health records, classification tools and language translators. The Metathesaurus, the Semantic Network and the SPECIALIST Lexicon are three different categories of UMLS knowledge sources.

The Metathesaurus is a large database containing over one million biomedical and health related concepts which are retrieved from several thesauri, indexed biomedial literature, controlled vocabularies, code sets and classification systems. The Metathesaurus is classified by concept and it connects each concept to its alternative names and views in other source vocabularies, it also determines appropriate relationships among concepts. The major semantic groups are organisms, anatomical structures, biologic function, chemicals, physical objects, and concepts or ideas [18].

To conduct the indexing process of our dataset we use MetaMap [2]. MetaMap is a program developed by National Library of Medicine (NLM), which maps biomedical texts to the UMLS Metathesaurus. MetaMap locates all the UMLS concepts associated with terms in biomedical texts.

3.2   Query Context Modeling

The medical query context reflects the most related concepts to the query; these concepts could represent symptoms, procedures or correlated diseases that are in relation with the query concepts.

The main motivation for using association rule mining for query context modeling is to find meaningful associations between concepts, especially when different variations of the same concepts occur in the documents. For example given the concept "GERD", Gastric Acid, Gastroesophageal reflux disease and GERD are all mapped to the same UMLS concept ID, "C0017168". Applying AR mining allows finding association between these concepts. Here, we consider a concept-based representation of the query obtained by mapping the query terms to the UMLS Methathesaurus.

Modeling a semantic-based query context consists the following steps: (1) extracting the representative concepts using association rule mining and (2) measuring semantic relatedness of the extracted concepts to query context for weighting the concepts in query context. Figure 1 presents our semantic-based query context-modeling algorithm.

3.2.1   Query Concepts Extraction Using Association Rule Mining

Query concepts extraction consists of extracting the rules that satisfy the query concepts, and weighting and ranking extracted concepts according to their semantic similarity/relatedness to the query concepts.

For each query, query concepts extraction consists of the following steps:

(1) A basic term-based retrieval model is used to retrieve n initial search results.
(2) Extract a list of rules by applying TopKRules algorithm on the n initial search results.
(3) Select the consequent of those rules that cover the query concepts in their antecedents.

In Fig.1. query concepts extraction corresponds to lines 1-6. For each query we extract, a list of rules L generated from TOPKRules algorithm, where LHS represents left hand side of the rule and RHS represents right hand side of the rule, the original query context Q_c, list of permutations of query concepts P, that contains all of the ordered combinations of query concepts and C, which is a series of concepts extracted via TopKRules. In order to extract concepts we compare left hand side of each rule in L to each of the permutations of the query concepts and save the right hand side of those rules covering any permutations.

INPUT: List of rules generated from TopKRules L, List of original query concepts Q_c, List of permutations of query concepts P, List of consequent concepts that cover query concepts C,

OUTPUT: List of concepts semantically weighted in relation to query concepts CTX_q,

1. **For** each item c in P
2.   **For** each item m in L
3.   If LHS$(L_m) = P_c$
4.   SAVE (RHS(L_n),C) ;
5.   **End for**
6. **End for**
7. **For** each item r in C
8.   SAVE (Semantic Similarity (C_r, Q_c), r, CTX_q);
9. **End for**

Fig. 1. Semantic-based Query Context Modeling

3.2.2   Query Concepts Weighting

Among the extracted concepts many are trivial, redundant or semantically wrong. In order to identify the relevant concepts we use semantic relatedness/similarity measures to discriminate the concepts with respect to the queries [17].

Semantic relatedness and semantic similarity are useful measures for effective natural language processing, artificial intelligence and information retrieval in medical domain [15]. Similarity measures determine how similar two concepts are by calculating how close they are in the UMLS semantic network hierarchy.   Relatedness measures determine how related two concepts are by using concepts definitions information.

For each concept in list of concepts that cover the original query concepts (C), we measure the concept's semantic similarity/relatedness to the query concepts (fig.1. lines 7-9). The output is $CTX_q = <C_1, C_2, C_3, ..., C_n>$, which is a medical query context represented by a series of concepts extracted via semantic-based AR mining.

For the purpose of this study we use the following measures:

Gloss Vector [21]: this relatedness measure represents concepts by vector of co-occurrences and measures relatedness of concepts by calculating cosine of vectors.

Wu and palmer [23]: this similarity measure calculates the similarity of two concepts by taking into account depth of two concepts in UMLS semantic network as well as Least Common Subsumer (sharing ancestors in the hierarchy).

3.3   Re-ranking Clinical Documents Using the Query Context

The re-ranking process is performed by integrating the semantic-based query context CTX_q into the information retrieval process. For each document d_k retrieved, we combine the intial score S_i based on okapi BM25 ranking model [4,9] and the contextual score S_c based on cosine measure between the d_k and CTX_q. We combine S_i and S_c using a tuning parameter α to balance the impact of the original score S_i and the new score S_c calculated according to our method.

$$S (d_k) = \alpha * S_i(q, d_k) + (1-\alpha) * S_c (d_k, CTX_q) \qquad (1)$$

$$0 < \alpha < 1$$

The contextual score is computed using cosine similarity measure between the retrieved results and concepts obtained from association rules.

$$S_c (d_k, CTX_q) = cos (\vec{d_k}, \vec{CTX_q}) \qquad (2)$$

4   Experimental Settings

In our search experiments, we compare the standard retrieval performance using only the query (ignoring any query context) to the contextual search performed using the query context. The standard retrieval is based on the Okapi BM25 scoring formula, where we set b to 0.75, k1 to 1.2 and k3 to 8. We use Terrier [20] for indexing the dataset using only terms, only concepts or both terms and concepts. For query context modeling we conduct experiments to find the optimal value for the following parameters: (1) confidence level, (2) number of documents selected for AR mining and (3) the number of concepts used for query context representation. For document re-ranking using equation (1) we tune the parameter α in [0 1].

4.1   Dataset

The corpus that is used to develop and test our approach is provided within the context of the TREC Medical Records Track 2011 challenge and is composed of query set, patient records and relevance judgments.

The query set is developed by physicians and contains 34 topics. Each topic specifies a particular disease/condition set and a particular treatment/intervention set. For the purpose of this study queries are semantically indexed using Metamap. Assume we are looking for concepts associated with the query "Patient with hearing loss", MetaMap examines the input text and generates a ranked list of relevant candidate concepts for phrases "patients" and "with hearing loss" and selects the highest ranked concept/s which are "C0030705: Patients [Patients or Disabled Group]", "C0011053: hearing loss [Deafness]", "C0018772: hearing loss [Hearing loss, Partial]" and " C1384666: hearing loss [Hearing Impairment] ".

The patients' record is a set of de-identified electronic medical records that made available for research purposes through the university of Pittsburg. These documents are organized by visits. There is a total of 17,267 visits and 101,711 reports in which, each visit contains between 1 to 415 reports. Each document contains the following sources of information that can be used for the task:

- Checksum, which is the unique report id,
- Type, which is general descriptor of the report,
- Subtype, which is more precise descriptor of the report,
- Chief complaint, which describes the main symptom or reason for which the patient seeks treatment,
- Admit diagnosis (as ICD-9 code),
- Discharge diagnosis (es) (as ICD-9 code) and
  Report text which is in free text form and is the central part of each report describing symptoms, signs, diseases, family history, lab results and so on.

The original corpus is organized by individual reports and there is a many-to-one relationship between reports and visits, where a visit is an individual patient's single stay at hospital and is associated with a set of patients' records. The University of Pittsburgh provides a table of this mapping and a visit is used as the unit of retrieval in the track. For the purpose of this study we generate visit-based documents where each document is composed by concatenating the patient reports associated with it.

4.2   Relevance Judgments and Evaluation Criteria

Relevance judgments are binary and reflect whether a visit is relevant or not with respect to the query. These judgments are provided by TREC organizers.

Our runs are evaluated using TREC's official measures: bpref, R-prec and P10. The bpref measure is designed for situations where relevance judgments are known to be far from complete. R-Precision is the precision after R documents have been retrieved, where R is the number of relevant documents for the topic. P10 is average of precision at 10 documents retrieved.

5   Experimental Results

Our evaluation objectives consist of the following: (1) evaluate the performance of our approach compared to a baseline approach, (2) evaluate the effectiveness of AR mining for query context modeling on the retrieval performance (3) evaluate the impact of semantic relatedness measures for query context modeling on the retrieval performance and (4) compare our approach to a query expansion technique and a re-ranking using basic AR mining.

5.1   Parameter Tuning

Our main objective for these experiments is to tune the parameters in our approach and to investigate the impact of association rule mining and semantic similarity measure on the retrieval performance.

Confidence level: For all the runs minconf 0.6 is chosen and this choice relies on the fact that setting higher minconf results in missing some interesting rules and setting lower minconf results in huge number of rules, which may contain irrelevant and insignificant information.

Number of documents: In order to find the optimal number of documents used as transactions for AR mining algorithm, we conduct different runs to evaluate the impact of this parameter on retrieval performance. We vary the value for the number of documents in the following set {20,50,100,500,1000}. For each value we perform our contextual ranking using all concepts from TopKRules algorithm and use Gloss Vector measure for query context modeling and setting α=0 in equation (1) for re-ranking. Our evaluation results show that the optimal number of documents is top 20 documents per query.

Number of concepts: Fig.2. presents the impact of using top 5, top 10, top 20 and top 40 concepts on the retrieval performance. We use the top 20 visits and the Gloss Vector relatedness measure for query context modeling. The figure reveals that the best performance is achieved using top 10 concepts, therefore we believe that for the purpose of this study the optimal number of concepts in the query context is 10.

Alpha(α): In this experiment, we vary α in [0 1] in equation (1). For each value we perform our contextual re-ranking using top 20 visits and top 10 concepts and Gloss Vector for query context modeling. Results confirm that the optimal α is 0.1.

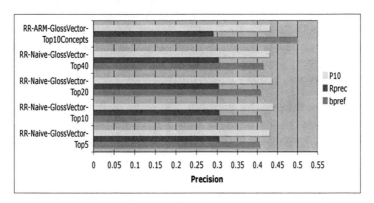

Fig. 2. Comparison of performance measures of using top 5, top 10, top 20 and top 40 concepts for query context modeling

5.2   Runs

Baseline runs:

- *Baseline-TermBased*: This run is based on the Okapi BM25 performed using term-based index.
- *Baseline-ConceptBased*: This run is based on the Okapi BM25 performed using concept-based index.
- *Baseline-Terms-Concepts*: This run is based on the Okapi BM25 performed using terms and concepts index.

Query expansion run:
QE-Terms-Concepts: this run is based on the Okapi BM25 using terms and concepts index where query expansion is performed using Rocchio's algorithm [22].

In the following runs we use only concepts of five semantic groups, which are DISO (Disorders), PROC (Procedures), PHYS (Physiology), CHEM (Chemical and Drugs) and ANAT (Anatomy) in order to obtain the most precise rules involving only medical related concepts and disregard general concepts such as CONC (Concepts and Ideas), GEOG (Geographic Areas) and LIVB (Living Beings). The following runs are conducted in order to evaluate the impact of semantic-based association rule mining on the retrieval performance (1) we compare our concept extraction using AR to a concept extraction using a naïve method for query context modeling and (2) we evaluate the impact of using semantic similarity/relatedness measures for weighting the concepts in the query context. The following runs are based on the re-ranking formula (equation (1)) with α =0.1 where the top 20 visits per query are used for query context modeling

Concept extraction using AR:

- *RR-ARM-Top20Trns-Top40Concepts*: this run relies on re-ranking the initial search results using query context that is composed of top 40 retrieved concepts (found as optimal value for this run) from TopKRules (k=1000) algorithm and weighted according to their support.
- *The RR-ARM-GlossVector-Top10Concepts*: this run relies on re-ranking the initial search results using query context that is composed of top 10 retrieved concepts from TopKRules (k=1000) algorithm and weighted according to their relatedness to query concept using Gloss Vector measure.

Concept extraction using naïve method: The naïve method for concept extraction consists of extracting all the concepts from top 20 documents for each query and weight and rank them according to their relatedness/similarity to query concepts.

- *RR-Naïve-GlossVector-Top10*: this run relies on re-ranking initial search results using top 10 concepts weighted according to their relatedness to query concepts using Gloss Vector measure.
- *RR-Naïve-WUP-Top10*: this run is relies on re-ranking initial search results using top 10 concepts weighted according to their similarity to query concepts using Wu and Palmer measure.

5.3   Performance Results and Analyses

Table 1 presents the retrieval performance of all the runs. The results show that our proposed approach (RR-ARM-GlossVector-Top10Concepts) outperforms all the baseline runs especially at the bpref measure. This proves the effectiveness of semantic query context modeling via association rule mining for improving the retrieval performance. Based on Table 1 we can draw the following conclusions:

- Baseline-Terms-Concepts run improves bpref by 19.5% and Rprec by18.5% compared to the Baseline-TermBased. This demonstrates the positive impact of using concepts for query and document representation, where it allows finding more relevant documents in the search results than using only terms.

Table 1. Performance Evaluations; Results are compared to Baseline-TermBased

Runs	bpref	R-prec	P10
Baseline-TermBased	0.3632	0.2606	0.4176
Baseline-ConceptBased	0.3126	0.2034	0.3242
Baseline-Terms-Concepts	0.4359	0.3084	0.4118
	(+19.5%)	(+18.5%)	(-1.3%)
QE-Terms-Concepts	0.4857	0.3336	0.4029
	(+33%)	**(+28%)**	(-3.5%)
RR-ARM-Top20Trns-Top40Concepts	0.4317	0.3030	0.4206
	(+18%)	(+16%)	(+0.7)
RR-Naïve-GlossVector-Top10	0.4086	0.3044	0.4382
	(+12%)	(+17%)	**(+5%)**
RR-Naïve-WUP-Top10	0.3940	0.2885	0.4281
	(+8.5)	(+11%)	(+2.5%)
RR- ARM-GlossVector-Top10Concepts	0.4985	0.2903	0.4300
	(+37%)	(+11%)	**(+5)**

- QE-Terms-Concepts run has been shown effective for improving the performance by 33% at bpref and 28% at R-prec, however the improvement of P10 has been negative.
- RR-ARM-Top20Trns-Top40Concepts run shows an improvement compared to the baseline runs but is not better than Baseline-Terms-Concepts run. We believe that this is due to the fact of ignoring the semantic relatedness of the concepts to the query.
- RR-ARM-GlossVector-Top10Concepts run outperforms term-based baseline run and improves bpref by %37, Rprec by %11 and P10 by 5%. This is due to the fact that this run takes into consideration both frequency of concepts and their semantic relatedness to the query concepts. This run also performs better than QE-Terms-Concepts run by improving bpref by %4 and P10 by 7%.
- RR-Naïve-GlossVector-Top10 run improves bpref by 12%, Rprec by 17% and P10 by 5% compared to the Baseline-TermBased. The improvement is due to taking semantic similarity into consideration even though a naïve method is used for concept extraction in the query context modeling. As Table 1 illustrates, Gloss vector relatedness measure outperforms Wu and Palmer similarity measure due to weakness of similarity measures in measuring closeness of concepts that are not connected in a tree hierarchy or they are located in different ontologies.

6   Conclusions and Future Work

In this paper we present a novel medical query context modeling based on association rule mining and semantic relatedness measures, which is then exploited for re-ranking clinical search results. Queries and documents are mapped to their correspondent medical concepts in the UMLS Metathesaurus. Query context modeling consists of extracting concepts from rules generated by AR mining algorithm. The extracted concepts are then weighted according to their semantic relatedness to the query concepts. We evaluate our proposed approach on the challenging ad-hoc retrieval task of TREC Medical Records Track. Results show that our proposed method improves the retrieval performance compared to the baseline search and two well-known methods, namely query expansion using Rocchio algorithm and naïve methods used for query context modeling.

Future work will focus on evaluating the impact of identifying temporal patterns using sequential pattern mining and time series mining on the retrieval performance. Also we plan to apply other semantic similarity and relatedness measures for improving clinical text retrieval performance.

References

1. Agrawal, R., Srikant, R.: Fast Algorithms for Mining Association Rules in Large Databases. In: Proc. of the 20th International Conference on Very Large Databases, pp. 487–499. Morgan Kaufmann, Santiago de Chile (1994)
2. Aronson, A.R., Lang, F.M.: An Overview of MetaMap: Historical Perspective and Recent Advances. J. Am. Med. Inform. Assoc. 17(3), 229–236 (2010)
3. Babashzadeh, A., Raissi, H., Daoud, M., Huang, X.J.: Query Expansion Using Multiple External Resources and Domain Knowledge for Medical Information Retrieval. In: Proc. of the 2012 Advances in Health Informatics Conference (AHIC 2012), Toronto (2012)
4. Beaulieu, M.M., Gatford, M., Huang, X., Robertson, S.E., Walker, S., Williams, P.: Okapi at TREC-5. In: Proc. of the 5th Text Retrieval Conference, National Institute of Standards and Technology (NIST), pp. 500–238. NIST Special Publication, Gaithersburg (1997)
5. Daoud, M., Kasperowicz, D., Miao, J., Huang, J.: York University at TREC 2011, Medical Record Track. In: Proc. of the 20th TREC, Gaithersburg, Maryland (2011)
6. Eichmann, D.: Concept-centric Indexing and Retrieval on Medical Text, Medical Record Track. In: Proc. of the 20th TREC, Gaithersburg, Maryland (2011)
7. Fournier-Viger, P., Wu, C.W., Tseng, V.: Mining Top-K Association Rules. In: Kosseim, L., Inkpen, D. (eds.) Canadian AI 2012. LNCS (LNAI), vol. 7310, pp. 61–73. Springer, Heidelberg (2012)
8. Huang, X.J., An, A., Hu, Q.: Medical Search and Classification Tools for Recommendation. In: Proc. of the 33rd International ACM SIGIR Conference on Research and Development in Information Retrieval, vol. 707. ACM, Geneva (2010)
9. Huang, X., Wen, M., An, A., Huang, Y.R.: A Platform for Okapi-based Contextual Information Retrieval. In: Proc. of the 29th Annual International ACM SIGIR Conference on Research and Development in Information Retrieval, Seattle, USA, pp. 728–728 (2006)
10. Icev, A., Ruiz, C., Ryder, E.F.: Distance-enhanced Association rules for Gene Expression. In: Proc. of the 3rd ACM SIGKDD Workshop on Data Mining in Bioinformatics, Washington, pp. 34–40 (2003)
11. Jie, W., Stepane, B., Beng Chin, O.: Mining Term Association Rules for Automatic Global Query Expansion: Methodology and Preliminary Results. In: Proc. of the 1st International Conference on Web Information Systems Engineering, vol. 366, IEEE computer society, Hong Kong (2000)

12. Koopman, B., Bruza, P., Sitbon, L., Lawley, M.: AHERC and QUT at TREC 2011 Medical Track: A Concept-based Information Retrieval Approach. In: Proc. of the 20th TREC, Gaitherburg, Maryland, USA (2011)

13. Latiri, C.C., Yahia, S.B., Chevallet, J.P., Jaoua, A.: Query Expansion using Fuzzy Association Rules between Terms. In: Proc. of the 2003 4th JIM International Conference on Knowledge Discovery and Discrete Mathematics. Mets, France (2003)

14. Liu, B., Hsu, W., Ma, Y.: Integrating Classification and Association Rule Mining. In: Proc. of the 4th International Conference on Knowledge Discovery and Data mining, pp. 80–86. AAAI Press (1998)

15. Liu, Y., McInnes, B., Meaux, G., Pakhomov, S.: Semantic Relatedness Study using Second Order Co-occurrence Vectors Computed from Biomedical Corpora, UMLS and Wordnet. In: Proc. of ACM International Health Informatics Symposium, IHI 2012, pp. 363–372. ACM press, Miami (2012)

16. McCormick, T.H., Rudin, C., Madigan, D.: Bayesian Hierarchical Rule Modeling for Predicting Medical Conditions. The Annals of Applied Statistics 6, 652–668 (2012)

17. McInnes, B.T., Pedersen, T., Pakhomov, S.V.: UMLS-Interface and UMLS-Similarity: Open Source Software for Measuring Paths and Semantic Similarity. In: Proc. of the American Medical Informatics Association Symposium, San Fransico (November 2009)

18. National Center for Biotechnology Information, http://www.ncbi.nlm.nih.gov

19. Ordonez, C.: Comparing Association rules and Decision Trees for Disease Prediction. In: Proc. of the International Workshop on Healthcare Information and Knowledge Management, pp. 17–24. ACM, Virginia (2006)

20. Ounis, I., Amati, G., Plachouras, V., He, B., Macdonald, C., Lioma, C.: Terrier: A High Performance and Scalable Information Retrieval Platform. In: Proc. of ACM SIGIR 2006 Workshop on Open Source Information Retrieval, Seattle, USA (2006)

21. Patwardhan, S., Pedersen, T.: Using WordNet-based Context Vectors to Estimate The Semantic Relatedness of Concepts. In: Proc. of the EACL 2006 Workshop on Making Sense of Sense: Bringing Computational Linguistics and Psycholinguistics Together, Trento, Italy, pp. 1–8 (2006)

22. Rocchio, J.J.: Relevance Feedback in Information Retrieval. In: Salton, G. (ed.) The SMART Retrieval System: Experiments in Automatic Document Processing, pp. 313–323 (1971)

23. Wu, Z., Palmer, M.: Verb Semantics and Lexical Selection. In: Proc. of the 32nd Annual Meeting of the Association for Computational Linguistics, Las Cruces, pp. 133–138 (1994)

24. Yin, X., Huang, J.X., Li, Z.: Mining and Modeling Linkage Information from Citation Context for Improving Biomedical Literature Retrieval. Information Processing and Management: An International Journal (IPM) 47(1), 53–67 (2011)

25. Yin, X., Huang, X., Hu, Q., Li, Z.: Boosting Biomedical Information Retrieval Performance through Citation Graph: An Empirical Study. In: Theeramunkong, T., Kijsirikul, B., Cercone, N., Ho, T.-B. (eds.) PAKDD 2009. LNCS (LNAI), vol. 5476, pp. 949–956. Springer, Heidelberg (2009)

26. Yin, X., Huang, X., Li, Z.: Promoting Ranking Diversity for Biomedical Information Retrieval Using Wikipedia. In: Gurrin, C., He, Y., Kazai, G., Kruschwitz, U., Little, S., Roelleke, T., Rüger, S., van Rijsbergen, K. (eds.) ECIR 2010. LNCS, vol. 5993, pp. 495–507. Springer, Heidelberg (2010)

A Comparison of Mobile Patient Monitoring Systems

Gaurav Paliwal and Arvind W. Kiwelekar

Department of Computer Engineering
Dr. B.A. Technological University, Lonere, 402103
Raigad (M.S.), India
{gvpaliwal,awk}@dbatu.ac.in

Abstract. A survey of Mobile Patient Monitoring Systems is presented in this paper. Mobile patient monitoring systems have shown progress in terms of basic core functionalities needed for monitoring and detection of biosignals. However, their deployment is restricted to a small segment of health care delivery. The objective of this survey is to identify causes behind low deployment. We have selected twelve mobile patient monitoring systems and compared them against a set of functional and non-functional requirements.

1 Introduction

Mobile Patient Monitoring (MPM), a sub-area underneath M-Health, refers to continuous or frequent measurement and analysis of biosignals of patients by employing mobile computing and wireless communication technologies [1]. Mobile patient monitoring is one of the techniques to reduce health care delivery costs associated with traditional patient monitoring systems. Mobile patient monitoring systems primarily acquire biosignals and transmit them to the remote location where a doctor can monitor vital signs to detect any abnormality. Through the use of advanced features in current generation mobile networks, MPM systems now aim to provide more personalized care through wearable, portable and implantable systems. Thus MPM is emerging as an efficient method for chronic diseases management.

Despite the numerous benefits offered by MPM systems, a very low uptake has been observed globally [2] for these systems. In this paper, we have surveyed various MPM systems to identify the reasons behind low acceptance of these systems by end users. A set of functional and non-functional properties are identified to compare different MPM systems. Our paper identifies a set of non-functional requirements that are emerging as desirable features for next generation MPMs.

Rest of the paper is organized as follows: Section 2 presents an overview of mobile patient monitoring systems. Different types of requirements for MPMs and comparison of various MPMs against these requirements is described in Section 3. Section 4 discusses some of the MPMs and their specific features. Results of the comparison are summarized in Section 5.

G. Huang et al. (Eds.): HIS 2013, LNCS 7798, pp. 198–209, 2013.

2 Mobile Patient Monitoring Systems

Existing mobile patient monitoring systems vary in terms of types and numbers of features supported by them. This section presents an overview of twelve existing MPMs. These MPMs are selected on the basis of diversity of features implemented in them, techniques used for detection, communication, interpretation and analysis of biosignals.

1. *Intelligent Mobile Health Monitoring System (IMHMS)*[3] The main objectives of IMHMS system is to intelligently predict patients health status and to provide them feedback through mobile devices. The IMHMS system uses Wearable Wireless Body Area Network to collect data from patients. The system stores the results of patient's examination and treatment in a central database. One of the main features of this system is use of data mining techniques to extract relevant information from biosignals. IMHMS also provides a flexible, simple and user-friendly interface. The system needs improvement with respect to providing secured transmission of biosignals.

2. *MobiHealthcare System (MHCS)*[4]
   The MHCS is mainly designed for cardiac, hypertensive or sub-healthy patients. The system provides specifically designed sensors to collect physiological signals. It collectively processes spatially and temporally collected medical data. The system is deployed on a server with big data storage where data mining and visualization is done. Although the system is capable of detecting abnormalities in biosignals and cardiac phenomenon, it can be extended to calculate the risk factors for cardiovascular and chronic diseases with more powerful data mining solutions.

3. *Multi-Touch ECG Diagnostic Decision Support System (MTDDS)*[5]
   The system is specially designed for cardiac patients. The system is capable of providing remote mobile communication to speed up diagnostic decision making using multimodal analysis engine. The prototype is capable to display ECG in three dimensional multi-layers on a multi-touch mobile device. The system prototype makes assessment faster and provides better medication management; however the system needs more contents and features for decision support system.

4. *Wireless Intelligent Sensor System (WISS)*[6]
   The WISS is intended for real time personal stress monitoring. The system provides affordable health monitoring services by utilizing plug-and-play sensor units complying with the common industry standard. It predicts critical performance aspects and stress resistance of soldiers under extreme conditions. The distributed wireless intelligent sensor system provides low-level real-time signal processing results in only transmission of compressed results. The system is convenient for prolonged stress monitoring, stressful training and normal activity. The system makes use of custom designed short-range communication devices to reduce power consumption and to increase the security. The system can be extended to evaluate the psycho-physiological state of individual persons.

5. *The MobiHealth System (MHS)* [7] The main objective of MHS is to provide a highly customizable vital sign monitoring system based on next generation public wireless networks. The MobiHealth System is based on generic service functional architecture platform provisioned for ubiquitous healthcare services. The system can support not only sensors, but a broad range of body worn devices and actuators. The measured vital signs are transmitted live over public wireless networks to healthcare providers. The major merit of MHS is its trial results, conducted successfully in many countries. Trials revealed low bandwidth and higher data loses as the major shortcomings of the existing network infrastructure.

6. *Mobile Cardiac Wellness Application (MCWA)*[8] The MCWA is a mobile patient-centric, self-monitoring, symptom recognition and self-intervention system that supports chronic cardiac disease management. The systems consists of back-end data repository, data mining, knowledge discovery, knowledge evolution and knowledge processing system, providing clinical data collection, procedural collection, intervention planning, medical situational assessment and health status feedback for users. It utilizes patient information and evidence based nursing knowledge to offer real-time guidance. The systems architecture has been presented from three different viewpoints as; an informational view (utilize multiple sources of information to construct patient specific health assessments and wellness), an operational perspective (data collection, patient assessments, patient evaluation, intervention planning and execution) and an architectural design view.

7. *Personalized Heart monitoring (PHM)*[9] The PHM system is aimed to combine ubiquitous computing with the mobile health technology. The system uses wireless sensors and smartphones to monitor the wellbeing of high risk cardiac patients. Smartphone examine real-time ECG data and determine whether the person needs external help or not. The system also consists of a fall detector and a Global Positioning System (GPS). Depending on the situation the smart phone can automatically alert pre-assigned caregivers or call the ambulance. The major shortcoming of the system is smartphones small battery life, usually drains in eight hours when continuously connected to the ECG device.

8. *Tele-Health Care System (THC)*[10] The main purpose of THC system design is to continuously monitor the heart attack patients. The THC system provides continuous mobility to both the patient and doctor. It detects the changes in Heart rate as well as blood pressure of the patient in prior and gives a self - alert ring to the patient and also sending an alert Short Messaging Service (SMS) to the doctor and thus gains immediate medical attention, results in reduced critical level of patient. The THC system implemented Alert services successfully but no steps have been taken to prevent false alarming.

9. *Ubiquitous Mobile Health Monitoring System for Elderly (UMHMSE)* [11] The system prototype is mainly designed for the elderly patients. The UMHMSE system provides the remote monitoring of human vital signs, mobility and location for collecting, gathering and analyzing data from a

number of biosensors. The system uses logistic regression technique to mine data and predict health risk from the knowledge of patients mobility, location and biosignal sensor data. The system suffers from smartphones small battery life and false alarms.

10. *Phone-Based E-Health System (PBEHS)*[12] PBEHS is specifically designed to remotely monitor Obstructive Sleep Apnea Syndrome (OSAS) patients. The system introduces a separate micro-control unit for data processing which significantly improves the smartphone battery life. The system offers a detailed analysis of energy consumption and presents a number of solutions to reduce system power consumption. The power consumption results are improved by 11 hours as compared to old prototype and still has possibility for further improvement using adaptive sampling, feature selection, compression, encoding and load balancing.

11. *CardioSentinal (CS)*[13] The main goal of the CS system is to provide an on-demand 24-hour heart care and monitoring services for elderly and outpatients. It provides monitoring services through biosensors, small-range wireless communication, pervasive computing, cellular networks and modern data centers. The system implements machine learning classification algorithms in order to identify ECG deflection patterns and to support decision making. The biosignal measurements collected by the system lacks in precision (upto 96% in some cases) as compared to the professional measurements in hospital. The system needs improvement with respect to robustness, communication reliability, accuracy, energy consumption and security.

12. *Advanced Health and Disaster Aid Network (AID-N)* [14] The system is purposely designed for the technological advancement of emergency response community services at Mass Casualty Incidents. AID-N system provides Electronic triage tags with built-in pulse oximeter and GPS to estimate the patient triage level and provide emergency services to the victims based on the triage level. The AID-N system uses service oriented architecture (SOA) that has shared data models of disaster scenarios to support the exchange of data between heterogeneous systems. The system itself has been tested successfully but the practical usability requires changes in emergency response protocols.

These systems use different terminologies to describe various components of MPMs. In the rest of the paper, we will use component names from the generic architecture of Patient Monitoring System proposed by Pawar P. A. in [1]. Figure 1 shows the generic architecture of MPMs, broadly divided into two components named: Body Area Network (BAN) and a Back-End System (BESys).

The BAN is defined as a network of communicating devices worn on or around the body which is used to acquire health related data to provide mobile health services to the patient. The BAN consists of a Mobile base Unit (MBU) and a set of BAN devices such as sensors, actuators or other wearable devices used for medical purpose. The sensors may directly transmit the biosignals data to the MBU or do it via the Sensor Front-End (SFE). The BESys comprises of the back-end server which can be of two types: back-end server to which the MBU

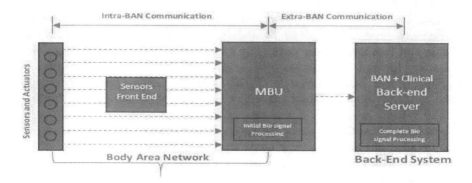

Fig. 1. Mobile Patient Monitoring System Generic Architecture

Table 1. Mobile Patient Monitoring Systems component describing Terminologies

System	Body Area Network(BAN)	Mobile Base Unit(MBU)	Back-end Server(BESys)
IMHMS	Wearable Body Sensor Network (WBSN)	Patients Personal Home Server (PPHS)	Intelligent Medical Server (IMS)
MHCS	Body Sensors	Mobile Device	Data and Data Mining Server
MTDDS	Sensors	Multi-Touch Smartphone	Web server
WISS	Wireless BAN of Intelligent Sensors (personal server (PS) and multiple WISE clients)	primary MBU functions are provided by BAN	Central Workstation
MCWA	Wearable Sensor Suite Mobile	Smart Phone	Server
PHM	Body Area Network	Mobile Base Unit	Back-end System
MHS	Body Area Network (BAN)	Mobile Base Unit (MBU)	back-end system
THC	Wrist Pressure Sensor	PIC Micro-controller	GSM MODEM (Mobile)
UMHMSE	Wireless Wearable Body Area Network (WWBAN)	Intelligent Central Node (ICN)	Intelligent Central Server (ICS)
PBEHS	Sensor Nodes And Micro-Control Unit (MCU)	Bluetooth Module Centralized Controller	Back-end Server
CS	BodyNets	Smartphone Gateway Nodes	Remote Data Centers
AID-N	Embedded Medical Systems for Triage and Biomedical Sensors	BESys is available at the scene in place of MBU	Ad Hoc Mesh Network

transmits biosignals data and clinical back-end server which may host custom health-care applications. Table 1 shows correspondences between components from the generic architecture and the components of the MPMs used for comparison.Though, the twelve surveyed system used different terminologies. The component interactions among them are similar to that of architecture specified above.

3   Comparison Framework

The requirements for MPMs capture the information necessary to build a MPM system from designer's and implementer's point of view. These are the set of precisely stated properties or constraints that a system must satisfy. The requirements of MPMs can be classified into two categories *functional* and *non-functional* requirements as given below:

Fig. 2. Representing Functional Requirements as Use cases

3.1 Functional Requirements

Functional requirements typically capture the core functionalities provided by a software system. There functionalities are performed either by a component of software system or an external agent. These functional requirements are represented through use case diagram as shown in Figure 2.

   i *Biosignal Processing(FR1)* The system should process biosignals and should be able to take decisions accordingly. Biosignals are initially processed by MBU on the basis of thresholds generated by the BESys and delivered to Back-End Server for further processing.

   ii *Biosignals Delivery(FR2)* First, Biosignals acquired by the BAN should be delivered to the MBU in real time. The communication between BAN and MBU is called intra-BAN communication. Second, Data should be delivered to back-end server instantly by MBU. The communication between the MBU and BE-sys is called extra-BAN communication.

   iii *Raise Emergency Alarm(FR3)* MBU and Back-end Server should collectively generate an emergency alarm in critical conditions, since the biosignals are processed by both the components.

   iv *Biosignals Interpretation(FR4)* The system should be able to diagnose the critical condition signs from biosignal. Biosignals must be correctly interpreted by MBU and Back-end Server.

   v *Biosignal Differentiation(FR5)* MBU should automatically discover the correlations between variations in physiological signals and lifestyle such as current activity, food intake and exercise.

   vi *Data Requisition(FR6)* To diagnose the patients current health status doctor or clinician needs the current biosignals as well as the past records from the database. The Back-End Server should be able to provide relevant data on request.

   vii *Communication(FR7)* The system should provide a communication interface between the patient and doctor. Graphical user interface provided on the MBU should felicitate user with an interface where he/she can interact with the doctor.

   viii *Medicine Infusion(FR8)* Sensors should be able to infuse the medicine into the patient body whenever triggered by the doctor or clinician. The requirement should be fulfilled by the BAN.

3.2   Non-Functional Requirements

i *Genericity (NFR1)* The System should be modified according to the patient monitoring needs. It should not be specific to certain disease, group or community of people.

ii *Security (NFR2)* Connection between the Mobile Base Unit and server should be secure and authenticated.

iii *Unique Patient Identification (NFR3)* Patient should be provided with a patientID by which he/she can be globally uniquely identified.

iv *Interoperability (NFR4)* The application should support various specified devices.

v *Privacy (NFR5)* System should maintain the patient privacy by restricting the access to the patient records.

vi *Intelligence (NFR6)* System should be capable of taking decisions on the basis of past and current records.

vii *Availability (NFR7)* System Should be available 24 X 7 for the continuous monitoring.

viii *Response Time (NFR8)* System should be fast enough so that on time emergency services can be provided to the patient.

ix *Easy Wear-ability (NFR9)* Body Area Network should be small in size, easy to wear and convenient for the patient.

x *Graphical interface (NFR10)* 1. The system should provide a graphical interface to display biosignals on MBU and Back-end Server 2. It should provide a graphical interface where doctor and patient can interact with the system.

xi *Accuracy (NFR11)* The data delivered to server should be accurate.

xii *Data loses (NFR12)* The system should be able to overcome data loses introduced due to various noise sources on the communication media.

xiii *Standards (NFR13)* System should follow various Standards provided for data sharing to achieve interoperability between the systems.

The aim of this comparison is to identify similarities and difference among MPMs described in Section 2. We have compared these systems against the functional and non-functional requirements stated in Section 3 The symbol ($\sqrt{}$) in Table 2 indicates the fulfillment of the requirement whereas the symbol X specifies the unimplemented or unidentified requirement. Table 2 shows enormous variabilities in the requirement implementation of the MPMs. It can be noticed from the Table 2 that the requirements FR8, NFR5, NFR12 and NFR13 are often missed out most of the MPMs during implementation.

4   Related Work

Our paper has compared twelve mobile patient monitoring systems against various functional and non functional requirements. In this section, we review some of the efforts proposed earlier and which are not covered in the comparison but has significant impact on MPMs with respect to MPMs architecture, BAN, MBU and BESys design.

Table 2. MPMs Comparison on Functional and Non-Functional Requirements

MPM System	FR1	FR2	FR3	FR4	FR5	FR6	FR7	FR8	NFR1	NFR2	NFR3	NFR4	NFR5	NFR6	NFR7	NFR8	NFR9	NFR10	NFR11	NFR12	NFR13
IMHMS	√	X	√	√	√	√	√	X	√	√	√	√	X	√	X	√	√	√	X	X	X
MHCS	X	√	X	√	√	√	√	X	X	X	X	X	√	√	√	√	√	√	X	X	X
MTDDS	X	√	X	√	X	√	√	X	X	X	X	X	√	X	X	X	√	√	√	X	X
WISS	X	X	X	√	√	√	√	X	X	X	X	X	X	X	X	√	√	√	√	X	X
MCWA	√	√	X	√	√	√	√	X	X	X	X	X	X	√	X	√	√	√	X	X	X
PHM	√	√	√	√	√	√	√	X	X	X	X	X	X	√	X	√	√	√	√	X	X
MHS	X	√	X	√	X	√	√	X	√	X	X	X	X	X	X	√	√	√	√	X	X
THC	√	√	√	√	X	√	√	X	X	X	X	X	√	X	√	√	√	√	X	X	X
UMHMSE	√	√	√	√	X	√	√	X	X	X	X	X	X	√	X	√	√	√	√	X	X
PBEHS	X	√	X	√	X	√	√	X	X	X	X	X	X	√	X	√	√	√	√	X	X
CS	√	√	√	√	√	√	√	X	X	X	X	X	X	√	X	√	√	√	X	X	X
AID-N	√	√	√	√	X	√	√	X	X	X	√	X	X	√	X	√	√	√	X	X	X

Most of the MPMs are designed for a specific group or community of people that are suffering from cardiovascular disease [8], Depressive Illness [15], dementia [16], hypertension [4], diabetes [17] or stress and some systems are dedicated to the older age group patients [11] and Mass Casualty Incidents [14]. Only a few architectures are presented for generic mobile patient monitoring systems [18] [19].

The Body Area Networks with some unconventional sensors sets like artificial endocrine pancreas [20], reflectance pulse oximeter [21], small range Bluetooth and an annular photo-detector to reduce power consumption have been offered. A mobile or PDA implementation for real time signal detection algorithm of patient ECG capturing and monitoring have been proposed in Mobile Health Monitoring Application Program [22].

In [23], Fei Hu et al. have identified networked embedded system design, network congestion reduction, and network loss compensation as three major performance issues for the MBU. To contend with such problems, the Personal Electrocardiogram Monitoring System described in [24] continuously monitors ECG and the system saves the ECG along with temperature into a smart phone. The system sends a multi media message to the base station with the current ECG image when it finds an irregular pattern in the ECG, whereas, the Mobile e-Health monitoring [25] shows a multi-tier agent based approach for MPM where the agent acts as a communication media between a patient and a doctor. System allows a patient to select whether the data is to be analyzed locally or on a centralized server. When data is analyzed locally it can prevent false alarms through interrogating a patient about his/her health status. These agents are normally deployed in mobile base units. The Remote Patient Monitoring System in [26] has realized secure and safe remote patient monitoring system that sends the monitoring data periodically to the clinical database. The clinical system automatically contacts the physician about any abnormality in the monitoring signals.

Technology has shifted its concern from individual systems to the distributed, cloud and grid systems where the computing power of the system has no limits but this shift has raised some new issues like interoperability, information sharing and data interpretation amongst different systems. The health information system [27] has emphasized on the need of the interoperability between the medical information systems that tries to provide interoperability between the heterogeneous medical information servers through service-oriented architecture [28] [29], web services and HL7 standards [30] [31].

As the patient monitoring systems are drifting towards maturity, researchers have tried to introduce intelligence into the system for the autonomous decision support. Intelligent patient monitoring system [32] where system can assist the physician in interpreting the medical data, decision making and automating the patient monitoring process through artificial intelligence and a data management system [33] which can effectively encapsulate, extract and interpret real world context aware information ensuring that physicians get the correct data every time. The data server reacts differently depending on the medical data and real time readings of the patients in different condition. To support these type of systems some techniques like data mining for predicting current health status [34] of a patient, applying clustering algorithm to both real time and historic data have been proposed. One more approach for simple data mining is presented as object oriented database system [35] for server and client.

The centralized server approach for data storage will open the doors to a gigantic biosignal data repository which can be used in various ways by research communities. On one hand, the centralized approach has provided various benefits whereas on the other hand, they raised certain issues with patient security and privacy. To deal with such issues the work of Johannes Barnickel et al. proposed to use AES-128 bit encryption algorithm for data storage at central server and password authentication [36] by a sensor server to access the monitoring data. Comprehensive Health Information System [37] has the capability to process patient data according to dynamically evolving set of data mining techniques and to share them among doctors, researchers and e-communities according to patient-defined access policies. Duke University's *Contain Explorer* [38] creates a view of administrative, financial, and clinical information generated during patient care by business intelligence tools from data warehouse.

Some papers presents a totally different aspect of MPMs like patients trajectory monitoring [39] to provide a better understanding of the effect of environmental factors on triggering health attacks in asthma patients hence support individual-based health care. WANDA [40] designed for early detection of Congestive Heart Failure symptoms and also provides feedback for regulating readings.

5   Conclusion

A framework to compare mobile patient monitoring systems with an objective to identify similarities and variabilities among existing MPMs is presented in this paper. It has been observed that most of the MPMs surveyed in this paper monitor biosignals specific to a disease such as patients suffering from cardiac problems. The core functionalities of MPMs include detection of biosignals, communication of biosignals, delivery of biosignals, and interpreting biosignals. Techniques to address the non-functional features such as secure transmission of biosignals, reduction in network congestion, reduction in power consumption, privacy of patient information have been supported by some of the existing MPMs. Other non-functional features such as interoperability between MPMs, autonomous monitoring system, extraction of relevant information from the monitored biosignals are emerging as critical design parameters for MPMs.

References

1. Pawar, P.A.: Context-aware vertical handover mechanisms for mobile patient monitoring. PhD thesis, Enschede, the Netherlands (October 2011)
2. Organization, W.H.: mhealth: New horizons for health through mobile technologies: second global survey on ehealth. Global Observatory for eHealth series, vol. 3 (2011)
3. Shahriyar, R., Bari, M., Kundu, G., Ahamed, S., Akbar, M.: Intelligent mobile health monitoring system (imhms). Electronic Healthcare, 5–12 (2010)
4. Miao, F., Miao, X., Shangguan, W., Li, Y.: Mobihealthcare system: Body sensor network based m-health system for healthcare application. E-Health Telecommunication Systems and Networks 1(1), 12–18 (2012)
5. Lin, M., Mula, J., Gururajan, R., Leis, J.: Development of a prototype multi-touch ecg diagnostic decision support system using mobile technology for monitoring cardiac patients at a distance. In: Proceedings of the 15th Pacific Asia Conference on Information Systems (PACIS 2011). Queensland University of Technology (2011)
6. Jovanov, E., O'Donnell Lords, A., Raskovic, D., Cox, P., Adhami, R., Andrasik, F.: Stress monitoring using a distributed wireless intelligent sensor system. Engineering in Medicine and Biology Magazine 22(3), 49–55 (2003)
7. van Halteren, A., Bults, R., Wac, K., Konstantas, D., Widya, I., Dokovski, N., Koprinkov, G., Jones, V., Herzog, R.: Mobile patient monitoring: The mobihealth system (2004)
8. Fortier, P., Viall, B.: Development of a mobile cardiac wellness application and integrated wearable sensor suite. In: SENSORCOMM 2011, The Fifth International Conference on Sensor Technologies and Applications, pp. 301–306 (2011)
9. Gay, V., Leijdekkers, P.: A health monitoring system using smart phones and wearable sensors'. Special Issue on'Smart Sensors in Smart Homes', IJARM 8(2) (2007)
10. Rajan, S., Sukanesh, R., Vijayprasath, S.: Design and development of mobile based smart tele-health care system for remote patients. European Journal of Scientific Research 70(1), 148–158 (2012)
11. Bourouis, A., Feham, M., Bouchachia, A.: Ubiquitous mobile health monitoring system for elderly (umhmse). arXiv preprint arXiv:1107.3695 (2011)
12. Gao, R., Yang, L., Wu, X., Wang, T., Lu, S., Han, F.: A phone-based e-health system for osas and its energy issue. In: 2012 International Symposium on Information Technology in Medicine and Education (ITME), vol. 2, pp. 682–686. IEEE (2012)
13. Gao, M., Zhang, Q., Ni, L., Liu, Y., Tang, X.: Cardiosentinal: A 24-hour heart care and monitoring system. Journal of Computing Science and Engineering 6(1), 67–78 (2012)

14. Gao, T., Massey, T., Selavo, L., Crawford, D., Chen, B., Lorincz, K., Shnayder, V., Hauenstein, L., Dabiri, F., Jeng, J., et al.: The advanced health and disaster aid network: A light-weight wireless medical system for triage. IEEE Transactions on Biomedical Circuits and Systems 1(3), 203–216 (2007)

15. Dickerson, R.F., Gorlin, E.I., Stankovic, J.A.: Empath: a continuous remote emotional health monitoring system for depressive illness. In: Proceedings of the 2nd Conference on Wireless Health, WH 2011, pp. 5:1–5:10. ACM, NY (2011)

16. Wai, A., Fook, F., Jayachandran, M., Song, Z., Biswas, J., Nugent, C., Mulvenna, M., Lee, J., Yap, L.: Smart wireless continence management system for elderly with dementia. In: 10th International Conference on e-health Networking, Applications and Services, HealthCom 2008, pp. 33–34. IEEE (2008)

17. Logan, A.G., McIsaac, W.J., Tisler, A., Irvine, M.J., Saunders, A., Dunai, A., Rizo, C.A., Feig, D.S., Hamill, M., Trudel, M., Cafazzo, J.A.: Mobile phonebased remote patient monitoring system for management of hypertension in diabetic patients. American Journal of Hypertension 20(9), 942–948 (2007)

18. Jones, V., Gay, V., Leijdekkers, P.: Body sensor networks for mobile health monitoring: Experience in europe and australia. In: Fourth International Conference on Digital Society, ICDS 2010, pp. 204–209. IEEE Computer Society (2010)

19. Pawar, P., Jones, V., van Beijnum, B.-J.F., Hermens, H.: A framework for the comparison of mobile patient monitoring systems. Journal of Biomedical Informatics (March 2012)

20. Poon, C.C.Y., Liu, Q., Gao, H., Lin, W.H., Zhang, Y.T.: Wearable intelligent systems for e-health. JCSE 5(3), 246–256 (2011)

21. Mendelson, Y., Duckworth, R., Comtois, G.: A wearable reflectance pulse oximeter for remote physiological monitoring. In: 28th Annual International Conference of the IEEE Engineering in Medicine and Biology Society, EMBS 2006, pp. 912–915. IEEE (2006)

22. Lee, D., Rabbi, A., Choi, J., Fazel-Rezai, R.: Development of a mobile phone based e-health monitoring application. Development 3(3) (2012)

23. Hu, F., Xiao, Y., Hao, Q.: Congestion-aware, loss-resilient bio-monitoring sensor networking for mobile health applications. IEEE Journal on Selected Areas in Communications 27(4), 450–465 (2009)

24. Tahat, A.: Mobile messaging services-based personal electrocardiogram monitoring system. International Journal of Telemedicine and Applications 2009, 4 (2009)

25. Chan, V., Ray, P., Parameswaran, N.: Mobile e-health monitoring: an agent-based approach. Communications, IET 2(2), 223–230 (2008)

26. Hariton, A., Creu, M., Nia, L., Slcianu, M.: Database security on remote patient monitoring system. International Journal of Telemedicine and Applications, 9 (2011)

27. Plácido, G., Cunha, C., Morais, E.: A soa based architecture to promote ubiquity and interoperability among health information systems (2011)

28. Plácido, G., Cunha, C., Morais, E.: Promoting ubiquity and interoperability among health information systems using an soa based architecture. Journal of e-Health Management, 2165–9478 (2012)

29. Abousharkh, M., Mouftah, H.: Soa-driven sensor-based patient monitoring system with xmpp based event notification

30. Schmitt, L., Falck, T., Wartena, F., Simons, D.: Towards plug-and-play interoperability for wireless personal telehealth systems. In: 4th International Workshop on Wearable and Implantable Body Sensor Networks, pp. 257–263. Springer (2007)

31. De Toledo, P., Lalinde, W., Del Pozo, F., Thurber, D., Jimenez-Fernandez, S.: Interoperability of a mobile health care solution with electronic healthcare record systems. In: 28th Annual International Conference of the IEEE Engineering in Medicine and Biology Society, EMBS 2006, pp. 5214–5217. IEEE (2006)

32. Fotiadis, D., Likas, A., Protopappas, V.: Intelligent patient monitoring. Wiley Encyclopedia of Biomedical Engineering (2006)

33. ODonoghue, J., Herbert, J.: Data management system: A context aware architecture for pervasive patient monitoring. In: Proceedings of the 3rd International Conference on Smart Homes and Health Telematic (ICOST 2005), pp. 159–166 (2005)

34. Patil, D., Andhalkar, S., Gund, M., Agrawal, B., Biyani, R., Wadhai, V.: An adaptive parameter free data mining approach for healthcare application. International Journal 3 (2012)

35. Ranjan, R., Varma, S.: Object-oriented design for wireless sensor network assisted global patient care monitoring system. International Journal of Computer Applications 45(3), 8–15 (2012)

36. Barnickel, J., Karahan, H., Meyer, U.: Security and privacy for mobile electronic health monitoring and recording systems. In: 2010 IEEE International Symposium on a World of Wireless Mobile and Multimedia Networks (WoWMoM). IEEE (2010)

37. Delaunay, G., Albino, A., Muhlenbach, F., Maret, P., Lopez, G., Yamada, I., et al.: The comprehensive health information system: a platform for privacy-aware and social health monitoring. IADIS e-Health (2012)

38. Horvath, M., Winfield, S., Evans, S., Slopek, S., Shang, H., Ferranti, J.: The deduce guided query tool: Providing simplified access to clinical data for research and quality improvement. Journal of Biomedical Informatics 44(2), 266–276 (2011)

39. Alkobaisi, S., Bae, W., Narayanappa, S., Liu, C.: A novel health monitoring system using patient trajectory analysis: Challenges and opportunities. In: GEOProcessing 2012, The Fourth International Conference on Advanced Geographic Information Systems, Applications, and Services, pp. 147–151 (2012)

40. Suh, M., Chen, C., Woodbridge, J., Tu, M., Kim, J., Nahapetian, A., Evangelista, L., Sarrafzadeh, M.: A remote patient monitoring system for congestive heart failure. Journal of Medical Systems 35(5), 1165–1179 (2011)

IT-Enabled Health Care Delivery: Replacing Clinicians with Technology

Deborah Fitzsimmons[1], Anthony Wensley[2], Ian Graham [3], and Gail Mountain[4]

[1] School of Health Studies, University of Western Ontario, London, ON, Canada
dfitzsi4@uwo.ca
[2] Management Department & ICCIT, The University of Toronto, Mississauga, ON, Canada
anthony.wensley@utoronto.ca
[3] Canadian Institutes of Health Research, Ottawa, ON, Canada
Ian.Graham@cihr-irsc.gc.ca
[4] School of Health and Related Research, University of Sheffield, Sheffield, UK
g.a.mountain@sheffield.ac.uk

Abstract. This paper shows how a Knowledge To Action framework may provide a useful lens for identifying the key phases and decision points in the implementation of technology-supported clinical systems.

1 Introduction

When discharged following a period of care in hospital, few of us would expect the clinician who will provide our transitional care to arrive in a cardboard box accompanied by an installation engineer. This, however, is increasingly becoming the case for patients discharged from hospital with a range of conditions including Chronic Obstructive Pulmonary Disease (COPD). Patients are increasingly being provided with tele-health monitoring technology (TMT) as a replacement for home visits by specialist clinicians. It is argued that this innovative approach to healthcare service delivery can expand clinical services by enabling providers to deliver care to more users using the same level of resources or, where resources are diminishing, allow the service to maintain steady state. It has also been shown that care may be delivered more effectively as in the case of the provision of home dialysis to diabetic patients [1].

TMT is defined as the remote exchange of physiological data between a patient at home and clinical staff to assist in diagnosis and monitoring [2]. TMT includes a home unit and peripherals used to measure and monitor temperature, blood pressure or other vital signs for clinical review at a remote location. Data transmission is typically via telephone lines or wireless technology. The use of this home-based medical monitoring technology offers real benefits for patients with long term conditions as it removes the geographical restriction to health care delivery, potentially reducing inconvenience to the patient, allowing them to remain in familiar settings and reducing the probability of medical complications resulting from being physically present in a clinical setting. There is evidence that use of TMT potentially results in a reduction in mortality rates and reduces the frequency of subsequent hospital visits by patients

G. Huang et al. (Eds.): HIS 2013, LNCS 7798, pp. 210–222, 2013.

requiring emergency attention. In addition, the effective use of TMT could lead to significant cost savings [3 as they allow clinicians to safely and effectively manage a larger caseload than was previously possible [4]; allow for the more reliable identification of patients likely to require additional clinical interventions; manage workload; and target care more appropriately (ibid, [5]). However, TMT involves the integration of a variety of different types of information technology into a traditional clinical service creating a range of implementation and service management issues.

The UK health care provider organisation which is the focus of our research was tasked with expanding its programme from a very limited number of patient referrals to all patients discharged from hospital with a diagnosis of COPD [6] with a minimal increase in resources. The management team identified that they would be unable to achieve this without some form of innovation. Changing to a tele-health supported service delivery model was identified as the only way the organisation could meet the service needs in their catchment area.

Despite the cited benefits of TMT however, it is reported that "implementation has proven complex" and that there are difficulties in scaling up [7] resulting many organisations undertaking small pilot studies [8]. The transfer of research findings into practice has been described as "a slow and haphazard process" and that "for many reasons, research findings are not being taken up in practice" – a situation which results in "inefficient use of limited health care resources" [9]. Researchers and decision makers are normally members of separate groups with distinct cultures and perspectives on research and knowledge, and it is acknowledged that neither group fully appreciates, acknowledges, or even understands the other's world [9]. Consequently, knowledge transfer in this context necessitates the bringing together of researchers and decision makers and the facilitation of knowledge transfer and integration. This knowledge transfer and integration requires knowledge translation to ensure that the resulting knowledge is both relevant and applicable for the decision makers and useful to the researchers.

A knowledge-to-action (KTA) framework [9] provides a conceptual map for representing the exchange, translation and merging of knowledge. We have applied one KTA framework to the implementation of a clinical service delivery system supported by TMT in the UK. We suggest that the KTA framework may provide a useful lens for identifying the key phases and decision points in the implementation of technology-supported clinical systems.

2   Methodology

In 2008, the National Institute for Health Research created nine Collaborations for Leadership in Applied Health Research and Care (CLAHRCs) to undertake high-quality applied health research focused on the needs of patients and addressing the gap identified by Cooksey [10] by supporting the translation of research evidence into practice in the UK NHS. The South Yorkshire CLAHRC incorporates a Tele health and Care Technologies (TaCT) group which undertakes research into the use of technologies to support self-management, self-efficacy, independence and well-being, focusing on technologies that collect and transfer data to health care professionals to support assessment and self-management of long term conditions in the community. TaCT was invited by the primary care services commissioner for one region within South Yorkshire to evaluate the effectiveness of a tele-health intervention designed to supporting patients discharged from hospital with early stage COPD.

Following the implementation of the new service and a period of stabilization, the researchers undertook a randomized controlled trial (RCT) of the service to examine its' impact upon hospital admissions and other health care resources, including emergency room visits; whether the programme helps patients to better manage their condition; and to determine whether it has an impact upon quality of life for the patient when compared to the traditional nursing model. This paper was developed based upon the observations of the researchers during the pre-trial planning phase.

3   Background

COPD is the fifth biggest killer disease in the UK [11]. It is the second most common cause of emergency admission to hospital and one of the most costly in-patient conditions [12], accounting for £587m of the £1.08bn spent on hospital admissions for lung disease by the NHS [13]. The current prevalence of COPD in the UK is around 1.5% of the population (approximately 900,000 people) [14]. COPD is a long term condition and is characterized by a chronic, progressive decline in lung function. As the disease progresses patients may become house-bound, socially isolated and depressed [15], thus experiencing a poor quality of life with impaired emotional, social and physical functioning (ibid). Exacerbations in symptoms occur with increasing frequency and often require hospital management.

On discharge from the local hospital, patients with early stage COPD could participate in an eight week programme during which they would receive six home visits by specialist COPD nurses or physiotherapists who would ensure the patient had safely transitioned from the hospital into their home environment, assess the patient's condition, provide education and develop a plan for self-management to help the patient to understand and cope with their condition. After eight weeks, if the patient was considered clinically stable, they would be discharged from the programme.

In May 2010, a tele-health supported service was introduced to enable the provider to extend their early discharge service to all patients discharged from the local hospital as they recognised that within their current resource level this would not be possible using their existing care pathway. In the new service, clinicians would only be required to undertake three clinical visits to confirm safe transition from hospital, collect a detailed medical history and provide a self-management plan. Professional installers then visited the patient's home and fitted a small base unit and various peripheral items to record basic nursing observations such as pulse oximetry. Over the eight weeks of the service, patients are required to use the system each day, answering a series of questions about their physical, social and emotional well-being which, along with their clinical data, is transmitted to a secure website via a secure server. Using a "traffic light" approach, the system compares submitted data with patient-specific parameters, triggering a system alert should they fall outside a key parameter or if the patient fails to provide data. Thus, community clinicians only needed to make additional patient visits to the three planned on the care pathway if an alert is raised.

4   Tele-health Implementation

The implementation of tele-health supported clinical services, by their nature, necessitates the development of service delivery processes that involve both information technology (IT) and clinicians. In the traditional service delivery mode clinicians interact with patients, monitoring them directly and operate in a familiar setting with respect to both location and power relationship. In contrast, the integration of TMT into the service delivery process results in the use of TMT to elicit relevant information from patients and the provision of clinical 'observations'. Consequently, it is understandable that clinicians may be concerned about whether the 'new' service delivery process will be as reliable or effective as the traditional service. Clinicians are trained to provide hands-on care and it is therefore reasonable to expect them to be more comfortable with and accepting of service delivery processes that involve face-to-face examination of the patient rather than reviewing patient data provided remotely by IT. Similarly, it is appropriate to question whether clinicians are likely to want to take control of the IT at the heart of the service delivery process, changing questions for specific patients or updating parameters as their condition changes. It is possible that clinicians would consider such activities as the responsibility of information systems (IS) professionals. In contrast, IS professionals can, and do, see the benefits that a technology system can bring and are familiar with the technological problems that can be encountered during installation but may feel uncomfortable with taking responsibility for what they consider medical activities. Thus, the key question is how individuals having distinctly different perspectives, knowledge sets, and comfort levels with respect to specific behaviours and interactions can integrate their knowledge and adopt modified roles and behaviours so as to actively engage in the new hybrid service delivery process and thus enable the health care provider meet the organizational challenges they face.

4.1   What Is Tele-health and Who Should Implement It?

Generally, it would seem natural that the implementation of a new information system would fall within the remit of the IT department of an organisation. Paradoxically, one would also expect clinicians to plan for, and implement, new processes and the clinical equipment associated with them. Thus, tele-health monitoring systems do not naturally fall within the responsibility, knowledge or competence of either of these two groups. This state of affairs in our view is one of the reasons why the implementation of such hybrid service delivery systems is problematic and challenging to all those concerned

From a technological perspective the base station installed in the patient's home acts as a dumb terminal, merely relaying data collected from the patient. It is necessary to provide a link to a centralised web-based system that requires strict multi-layer access controls to protect patient data; a link to the National NHS (or N3) Network to obtain fast broadband networking services for data sharing with maximum efficiency yet minimal risk; and the need to support customization to allow the system to be tailored to meet the requirements for each patient. As we have noted above, the expertise to implement and support such a system would seem to lie with the IT professionals who could work easily with the system developers and retailers, however, in the PCT installation this was not the case as we will discuss shortly.

We would note that clinicians are often involved in the implementation of new clinical devices [16] and electronic forums exist to facilitate this [17]. Clinicians understand issues such as clinical care pathways and infection control which are unlikely to resonate with IT professionals. Lack of attention to these, however, creates risks for the patient. It would appear that this consideration along with the belief that items that will be used by a patient as part of their care should be managed by a clinician that appears to have established that the responsibility for the implementation of clinical devices rests firmly within the clinical sphere of management at the PCT.

However, the question remains, who should be involved in the implementation and management of the hybrid systems such as tele-health service delivery systems? In developing their project plan for TMT implementation, the PCT realized that they did not have the resources or expertise to undertake the installation of equipment into patient's homes. Within the city, the Local Authority (LA) had established a service providing a 24-hour, 365 day a year emergency response service for approximately 7,000 residents of the Borough. Given their experience in installing comparable technology into people's homes the PCT created a partnership with the LA, tasking them with the installation of the tele-health equipment within 72 hours of hospital discharge: an innovative model requiring detailed planning and a great deal of collaboration.

Management of the implementation was undertaken by a project manager from the LA partner, and a management team comprising of the clinicians delivering the service, the service commissioner, a LA senior manager, and the researchers. Interestingly, the IT department were not invited to be members of the management team but were asked to undertake specific IT tasks including establishing the N3 [18] secure high speed network connections.

Both operational partners provided extensive support to the project, developing detailed flowcharts to identify key activities, responsibilities and timings to ensure all elements of the project remain on track; developing training materials for staff and patients; shared forms to capture data required by all partners and shared spreadsheets to track patient-specific data to ensure there are no gaps in service delivery.

5   Knowledge to Action Framework Cycle

Graham et al. (2006) developed a conceptual map for the KTA process, suggesting that knowledge creation and utilisation can be modelled as two separate processes that may be undertaken by different groups, possibly independently of each other, and at different points in time [9]. It is also possible that knowledge creation and application can be undertaken by researchers and knowledge users collaboratively.

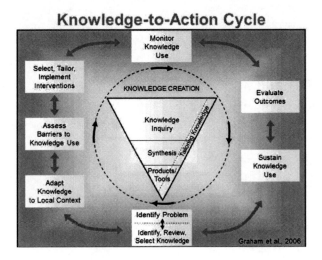

Fig. 1. Pictorial representation of the Knowledge-to-Action Framework

As shown in Figure 1 above, the knowledge creation process is represented as a funnel [9] whereby a multitude of unrefined, primary research studies are aggregated through the application of explicit, reproducible sense-making methods resulting in knowledge. These typically take the form of systematic review including meta-analysis and meta-synthesis. This more refined knowledge can be translated into knowledge tools such as guidelines or care pathways which can be used by stakeholders, thereby facilitating the uptake and application of knowledge.

If knowledge creators work with knowledge implementers, it becomes possible for researchers to tailor their research and dissemination methods to those best suited to address the problems identified by those best placed to implement the knowledge, thereby creating greater opportunity for adoption of research into practice. This collaborative approach has been referred to as participatory research, mode two knowledge production and integrated knowledge translation research [19].

6   Application of the KTA Cycle

Working with the partners and the management team during the implementation and stabilization phases of this initiative, the researchers realized that they were playing a vital, but subconscious role: they were facilitating the KTA cycle for the program partners and the project management team.

Fig. 2. Diagrammatic representation of the service partners

6.1   Knowledge Creation

When considering the knowledge creation funnel, the researchers undertook tasks at all three layers: a primary study of the standard clinical care service to determine the degree of consistency of delivery; a modelling exercise of the service to determine whether the newly developed clinical care pathway was being followed and to identify whether there were issues in service delivery that needed to be addressed prior to undertaking the RCT; and they synthesized existing knowledge from a range of published studies, filtering the information to identify what may assist or enhance the new service, feeding this to the PCT in the form of useful tools.

6.2   Action Cycle Phase One

The action cycle of knowledge implementation can commence in one of two ways: through the identification and pursuit of the answer to an issue, or the identification of knowledge and determination that there is a knowledge-practice gap that needs to be addressed [9]. In this instance, the two forms of initiation converged: the PCT identified an impending service gap if they continued to use their standard treatment modality, whilst the commissioner identified that a tele-health supported service might be a cost-effective and efficient method for delivering community-based care but this would be a significant change in delivery which would require support.

6.3   Action Cycle Phase Two

In phase two (adapting the knowledge to the local context), the researchers identified and presented elements of published evidence about tele-health and comparable health technology implementations to the project management team for consideration and determination of the applicability to the planned local service. Overcoming organisational knowledge barriers [20] to develop an understanding as to how the technology can be integrated into the organisation is a time-consuming process. Hartswood et al. [21] suggest that users "need the opportunity that only their work can offer to explore fully the possibilities for adopting, and adapting to, new systems and artefacts", and that as a user becomes more experienced they "develop new ways of using the system that in turn generates ideas for its further development". In this case experience highlighted inadequacies of the system. To allow clinicians to respond to alerts triggered in the system during their working hours, patient data must be uploaded each morning. Many patients with COPD find their condition is worse in the morning and are frequently late risers. Questions asking how their symptoms have been that day when they may have only been out of bed a short time were confusing for the patient and led to data omissions. Consequently the clinicians reframed the questions, asking how symptoms had been since the patient last entered their data. This represents a relatively minor change but one that has been well received by the patients.

The researchers suggested the need for calibration of a sample of the TMT equipment based upon a study which identified that, when tested, an alarmingly high proportion of equipment provided by a supplier was incorrectly calibrated and had to be returned [22]. This would take time and add cost to the project, but would assure the PCT of the validity of the readings, ensuring all generated alerts were based on accurate data. As false alerts could trigger additional visits and add to the cost of operating the service, the PCT decided to add this into their service planning process.

Similarly, the researchers recommended that the equipment should be subject to initial and repeated Portable Appliance (PAT) testing to ensure the electrical safety of the equipment. Whilst an individual buying an electrical product for use in their own home could assume that the product had been tested and was safe to use, as the third party provider of the TMT, the PCT would assume liability for the safety of the equipment. Consequently, they would be required to ensure the equipment had undergone PAT testing and was considered safe in accordance with the Electricity at Work Regulations [23]. The project management team again accepted this suggestion and then needed to determine who could undertake this activity on behalf of the PCT, establish annual scheduling of testing for each item of equipment between installations in patient homes and, of course, pay the additional cost for this.

6.4   Action Cycle Phase Three

In phase three of the action cycle, barriers to the implementation of the knowledge that may impede or limit uptake of the knowledge should be identified so that these barriers may be targeted and hopefully overcome or mitigated by intervention strategies. Additionally, the barriers assessment should identify supports or facilitators that can be taken advantage of.

Several logistical barriers were identified by the LA partner, including the fact that the TMT was supplied with a 1.5m telephone cord and 1.5m power cable. From previous experience of installing tele-care systems, they identified that approximately 25% of homes would not have a telephone socket and electrical outlet within 3m of each other. They also identified that one telephone network provider could not support the use of tele-health equipment due to protocol signalling issues.

Once they started to review system data, the clinicians identified that the majority of patient alerts triggered in the first two weeks of the service were usually related to the health status and well-being of the patient, rather than clinical issues. Patients with early stage COPD have yet to come to terms with their new health status 'norm' and recognise that it will take them some time to recover. Seeing this in the data captured through the TMT highlighted this issue. Responding to the resultant increase in alerts increases clinician workload and can be a barrier to the operation and sustainability of tele-health supported services by reducing the availability of clinicians to care for additional patients and reducing clinician confidence in the system.

6.5   Action Cycle Phase Four

In phase four of the action cycle, interventions are selected, tailored and disseminated during the planning of implementation of the knowledge. When the PCT raised the cabling issue with their health and safety staff, they were informed that an extension cord for the power cable would constitute a 'trip and fall' hazard, leaving the PCT liable to law suits. This would only be acceptable if the cords were securely fastened to the wall to prevent such a mishap. However, this would clearly extend the time and the cost associated with installation and it was felt that users would be extremely unhappy about any permanent damage to their homes resulting from the installation of temporary equipment. The PCT determined that it was inappropriate to exclude patients from the use of potentially valuable assistive technology merely due to the cabling limitations of their homes. The researchers recommended using a 3m replacement telephone cord whose small diameter meant it could easily be fastened to the wall by means of a clip with self-adhesive backing. This solution has been incorporated into the installation process where necessary.

Unfortunately, no published solution to the telephony switching issue could be identified by the researchers. For long-term patient monitoring systems it may be necessary for the patient to switch telephone network provider if they wished to use a tele-health supported solution, but given the time taken to do this and the significant cost involved, this is not a viable solution for this short-duration service. Consequently, the management team decided that patients using that particular telephone service provider could only be offered the traditional clinical.

Having identified the nature of the alerts triggered during the first two weeks of service, clinical staff are now providing additional patient education about this aspect of the care pathway and the impact is being seen in fewer alerts during this 'settling in' period by new patients. This provides some indication as to the effectiveness of the KTA framework as an approach to improving service provision.

6.6   Action Cycle Phase Five

Limited monitoring of knowledge use has been conducted during the pilot trial. However in interviews conducted by the researchers, patients reported that they liked the technology and, contrary to clinician expectations, were comforted by the thought that clinicians were constantly monitoring their condition and would contact them should there be a potential problem with their health. Of the patients who received the TMT supported service and provided feedback, 100% found the equipment easy to use; 89% felt it helped them to manage their COPD; and there was a 92% compliance rate for completion of daily TMT assessments with all patients reporting that they were confident that there would be help available if their condition deteriorated. This served to reassure the clinicians who then became more committed to the project and has imbued the staff with greater confidence in the technology, demonstrating that TMT can safely be used as a replacement for home visits by specialist staff.

6.7   Action Cycle Phases Six and Seven

The pilot tele-health service has only superficially touched upon the final phases of the KTA cycle: evaluation of outcomes and sustainability which will be examined in depth during the definitive trial. Clearly, evaluation is critical to determine the effectiveness and utility of the KTA framework and to compare its efficacy with other approaches.

7   Discussion

In the National Center for the Dissemination of Disability Research review of KT models [24] it was identified that the KTA cycle presents a more comprehensive picture of KT as it incorporates the knowledge creation process and the need to adapt the knowledge to fit with the local context and sustain knowledge use by anticipating changes and adapting accordingly. It is assessed to be a comprehensive framework that begins to incorporate the full cycle of KT from knowledge creation through implementation and impact [24].

Facilitating the KTA cycle has been undertaken by the researchers whilst maintaining their independence and respecting their research brief. Where they identified knowledge that could be useful for the new service, this was presented with supporting evidence, but the researchers then allowed the management team to discuss the knowledge and make their own operational decisions. Where the project management team asked for information that could assist them in the resolution of issues, again this was presented by the researchers, but all operational decisions were taken by the project management team.

The clinicians identified where the tele-health supported service needs to be improved and have taken active steps to address the issues. This reinforces engagement and the clinicians are committed to the full implementation of this service, working towards that objective by developing clinical, administrative and evaluative processes designed to work when the system is fully deployed at maximum capacity rather than interim solutions that cannot be scaled up and require replacement in the future. Whilst IT professionals were involved in the implementation of an N3 network connection this was the limit of their involvement. The project management team took responsibility for the implementation of this complex system, effectively using an evidence-informed approach to guide their decision-making processes. The project management team accomplished this for themselves, gaining greater confidence in, and understanding of the system in the process. In effect, the TMT is doing as Berg [25] suggests and has formed "an artful integration" with the clinicians using it with "skilful interaction" to replace unnecessary visits whilst maintaining a high standard of patient care, which has supported the mainstream adoption of the technology in the region.

8   Conclusion

We believe that this implementation has demonstrated how the KTA cycle can provide a useful framework to support the integration of clinical service delivery with IS technology and enable a health care organisation to create a cost-effective and efficient service which will meet their service delivery mandate in the future. Whilst we have applied KTA framework to the implementation of an IT-enabled clinical service, we feel that this approach would be equally effective in any implementation where an evidence-informed approach to service delivery is required

Although Graham et al. [9] have provided a cognitive map that helps to explain and guide the process by which knowledge can be translated into practice, we feel that there is a need to develop the KTA framework further, applying it specifically to the implementation of complex clinical/informatics interfaces such as TMT. A clinical technology KTA framework could be used by researchers in the future to ensure that they address these needs in their research questions and in their approach to dissemination to ensure best practice is adopted in service delivery, and be developed into a user guide for health care managers to explain to them how the framework could be used to support implementation.

Acknowledgements. This research was supported by National Institute for Health Research Collaboration for Leadership in Applied Health Research and Care, South Yorkshire. Further details can be found at http://www.clahrc-sy.nihr.ac.uk.

References

1. Canadian Agency for Drugs and Technologies in Health,
   http://www.cadth.ca/media/pdf/H0475_Home_Telehealth_tre.pdf
2. Curry, R.G., Trejo Tinoco, M., Wardle, D.: http://www.telecare.org.uk/
   shared_asp_files/GFSR.asp?NodeID=46395
3. Audit Commission, http://www.auditcommission.gov.uk/
   SiteCollectionDocuments/AuditCommissionReports/NationalStudies/
   NationalReport_FINAL.pdf

4. Broderick, A.: http://www.techandaging.org/ARC_Presentation.pdf
5. Darkins, A., Ryan, P., Kobb, R., Foster, L., Edmonson, E., Wakefield, B., Lancaster, A.E.: Care Coordination/Home Telehealth: The Systematic Implementation of Health Informatics, Home Telehealth, and Disease Management to Support the Care of Veteran Patients with Chronic Conditions. Telemedicine and e-Health 14(10), 1118–1125 (2008)
6. Fitzsimmons, D.A., Thompson, J., Hawley, M., Mountain, G.A.: http://www.trialsjournal.com/content/12/1/6
7. European Commission, http://ec.europa.eu/information_society/ activities/einclusionlibrary/studies/docs/ict_ageing_ final_report.doc
8. Whole System Demonstrator Action Network, http://www.wsdactionnetwork.org.uk
9. Graham, I., Logan, J., Harrison, M., Strauss, S., Tetroe, J., Caswell, W., Robinson, N.: Lost in knowledge translation: Time for a map. J. Contin. Educ. Health. Prof. 26(1), 13–24 (2006)
10. Cooksey, D.: http://www.hm-treasury.gov.uk/d/ pbr06cookseyfinal_report_636.pdf
11. Royal College of Physicians of London, British Thoracic Society and British Lung Foundation, http://www.rcplondon.ac.uk/clinical-standards/ceeu/ Current-ork/ncrop/Documents/Report-of-The-National-COPD- Audit-2008-clinical-audit-of-COPD-exacerbations-admitted-to- acute-NHS-units-across-the-UK.pdf
12. Meldrum, R., Rawbone, A., Curran, D., Fishwick, D.: http://www.lunguk.org/ Resources/BritishLungFoundation/MigratedResources/Documents/ I/InvisibleLivesreport.pdf
13. British Lung Foundation, http://www.lunguk.org/Resources/ BritishLungFoundation/MigratedResources/Documents/L/ Lung_Report_III_WEB.pdf
14. British Lung Foundation, http://www.lunguk.org/Resources/ BritishLungFoundation/MigratedResources/Documents/L/ Lu_Report_III_WEB.pdf
15. Florence Sueng Kim, H., Kunik, M.E., Molinari, V.A., Hillman, S.L., LaLani, S., Orengo, C.A., Petersen, N.J., Nahas, Z., Goodnight-White, S.: Functional Impairment in COPD Patients: The Impact of Anxiety and Depression. Psychosomatics 41, 465–471 (2000)
16. Pennine Acute Hospitals NHS Trust, http://www.pat.nhs.uk/uploads/ 20080717_052%20RI%20scanner.pdf
17. EBME Ltd., http://www.ebme.co.uk/forums/ubbthreads.php/ forums/9/1/Medical_EquipmentManagement
18. Canadian Institutes of Health Research, http://www.cihr-irsc.gc.ca/e/ 39033.html
19. National Health Service, http://www.connectingforhealth.nhs.uk/systemsandservices/n3
20. Tanriverdi, H., Iacono, C.S.: Knowledge Barriers to Diffusion of Telemedicine. International Conference on Information Systems. In: Hirschheim, R., Newman, M., DeGross, J.I. (eds.) Proceedings of the International Conference on Information Systems, pp. 39–50. Association for Information Services, Atlanta (1998)
21. Hartswoods, M.J., Procter, R.N., Rouchy, P., Rouncefield, M., Slack, R., Voss, A.: Working IT Out in Medical Practice: IT Systems Design and Development as Co-Realisation. Methods Inf. Med. 4, 392–397 (2003)

22. Watson, J.M., Kang'ombe, A.R., Soares, M.O., Chuang, L., Worthy, G., Bland, J.M., Iglesias, C., Cullum, N., Torgerson, D., Nelson, E.A.: Use of weekly, low dose, high frequency ultrasound for hard to heal venous leg ulcers: the VenUS III randomised controlled trial. BMJ 2011 342, d1092 (2011)
23. HM Government,
    http://www.opsi.gov.uk/si/si1989/Uksi19890635_en_1.htm
24. National Center for the Dissemination of Disability Research,
    http://198.214.141.98/kt/products/ktintro/ktmodels.html
25. Berg, M.: Patient care information systems and health care work: a sociotechnical approach. Int. J. Med. Inform. 55, 87–101 (1999)

Instrument to Assess the Need of Disabled Persons for Rehabilitation Measures Based on the International Classification of Functioning, Disability and Health

Alexander Shoshmin, Natalia Lebedeva, and Yanina Besstrashnova

Federal State Institute "Saint Petersburg Scientific and Practical Center of Medical and Social Expertise, Prosthetics and Rehabilitation of the Disabled named after G.A. Albrecht", Saint Petersburg, Russia
{shoshminav,nn_lebedeva,besstjan}@mail.ru

Abstract. This paper introduces a special software product to assess the need of disabled persons for rehabilitation. It is based on a codifier of disability categories which are differentiated according to the primary type of assistance. Depending on impairments severity of body functions and structures and using the International Classification of Functioning, Disability and Health it allows to choose rehabilitation services and technical aids for rehabilitation when making the Individual Rehabilitation Program. The obtained condition codes of a disabled person can be used to electronically record measures on a social card of a citizen, to control the implementation of rehabilitation measures, to analyze the needs in various kinds of assistance and accordingly to plan a budget for different levels etc.

The software product has been developed in DBMS Caché in programming environment qWord by SP.ARM Company (St.Petersburg, Russia). The software is multilingual and can be adapted for almost any national language.

Keywords: software product for need assessment, disabled persons, rehabilitation, the International Classification of Functioning, Disability and Health.

1   Introduction

In the Russian Federation it is a duty of the state to render social support for the disabled. The Federal State Institution of Medical-Social Expertise (MSE) which has a widespread system of institutions around the country evaluates the need of the disabled for social support incl. rehabilitation. The Russian government admits that activity of MSE and the rehabilitation system of the disabled is not aimed to achieve common goals, as the result, the efficiency of rehabilitation measures remains low [1].

In order to raise the level of objectivity, accessibility and efficiency of expert and rehabilitation measures the Concept of Improving the State System of Medical-Social Expertise and Rehabilitation of the Disabled was developed and approved by the Russian government [2]. Implementation of the Concept shall provide the compliance of principles and measures of support of the disabled with the requirements of the UN Convention on the Rights of Persons with Disabilities [3-4].

G. Huang et al. (Eds.): HIS 2013, LNCS 7798, pp. 223–231, 2013.

A new variant of a single system for MSE is being developed and integrated. It is supposed to function on the basis of cloud computing and be used in all the subjects of the Russian Federation. The information system shall provide the evaluation of citizens' condition, determine measures and services raising the rehabilitation efficiency of the disabled and providing the information exchange with other organizations. Therefore in order to define the need of a disabled person for rehabilitation we have developed a special software product based on DBMS InterSystems Caché®.

Caché was chosen as "an advanced object database that provides in-memory speed with persistence, and the ability to handle huge volumes of transactional data. It can run SQL faster than relational databases. Caché enables rapid Web application development, massive scalability, and real-time queries against transactional data – with minimal maintenance and hardware requirements" [5].

The software product was developed in programming environment qWord-XML by SP.ARM Company (St.Petersburg, Russia). It supports a hierarchical data model, which allows to express semantics domain in a natural way. qWord-XML is a powerful tool for a developer to create input and output forms. The database schema is not declared explicitly, and may be changed during the maintenance and development of information systems. qWORD-XML helps to create both operational and analytical systems, combining features of the universal browser [6]. The software is multilingual and thus can be adapted for almost any national language. Development of this product required brand new ideology which had never been used in Russia before.

2   Basic Provisions of the Software Product

An important step for improving the medical-social expertise is to bring systems of statistic record, evaluation of disability, development of individual rehabilitation program of a disabled person and efficiency of its implementation in accord with international standards. The standard in this field is the International Classification of Functioning, Disability and Health (ICF) adopted by the WHO Expert Committee in 2001 [7]. In order to determine the selection of measures for social support for the disabled in different life situations we have developed a codifier of disability categories which is based on the ICF and differentiated according to the primary type of assistance a disabled person need (the Codifier). The codifier of disability categories is an instrument which allows to define the primary type of assistance depending on disabling impairments of body functions and structures and to define the type of situational assistance that a disabled person need depending on his body functions. Taking into consideration the Codifier, situational assistance shall be interpreted as a process or complex of measures intended to form relationships between a disabled person and the environment. The situational assistance shall be carried out in social institutions and with consideration of a letter code defined by the Codifier and the situation in which a disabled person is, such as: in the shop, public transport, administrative institutions etc.

The Codifier can be used assessing the condition of a disabled person at examination, developing the Individual Rehabilitation Program (IRP), evaluating the efficiency of its implementation, choosing technical aids for rehabilitation of disabled persons and certain measures of the IRP, and also to evaluate the efficiency of rehabilitation measures that are performed.

3   Information Structure of the Codifier

The Codifier is based on the main levels and parts of the ICF [7] and the national standard of the Russian Federation on the classification of technical aids for rehabilitation of disabled persons [8] (analog ISO 9999:2002), classification of rehabilitation measures and services.

It was required to create software for implementing to-date knowledge and approaches reflected in a number of the logically related but independent classifiers. This software product is aimed to practical usage in working with disabled people. Until that moment, most of these classifiers were used mainly for statistical purposes and never to identify priority assistance to help people in a real time.

The analysis showed initial attributes that defined the need for assistance. They proved to be body functions impairments and were mainly defined by body structures impairments. The same body functions impairment can be caused by disorders of different body structures. Depending on the structures impairments different types of assistance can be delivered (in case of the same function impairment). A function impairment leads to a change in activities and participation. To increase the human's inclusion to the society is possible by the impact to specific elements of activities and participation. Accordingly, the impact may be described in terms of technical aids of rehabilitation services. The Codifier's framework is based on all these assumptions.

The structure of the Codifier is the following: it consists of independent records (codes) linked in chains based on categories of the ICF domains with appropriate qualifiers, the National Standard Specification of the Russian Federation GOST R 51079-2006 (ISO 9999:2002) "Technical Aids for Persons with Disabilities. Classification" [8] and the classification of rehabilitation services.

We have singled out five levels of the Codifier: the first level – category of disability, the second level – body functions + qualifier of impairment, the third one – body structures + qualifier of impairment, the fourth one – activity (execution of a task or an action by a disabled person) and participation (involvement of a disabled person in a life situation) + appropriate qualifiers; the fifth level – technical aids of rehabilitation and rehabilitation services.

The first level of the Codifier (the category of disability) is qualified by impairment of body functions leading to certain categories of disability connected with primary type of assistance (permanent, partial and/or situational) that a disabled person need.

Each category of disability has its own letter code, which is written in the same way both in Russian and Latin alphabets:

A) Needs permanent outside care (assistance, supervision) because of severe restrictions in mobility, self-care, domestic life to and/or orientation.[1]

B) Needs partial outside care (assistance, supervision) incl. at home because of severe restriction of mobility.

C) Needs partial outside care (assistance, supervision) and guidance, incl. when outside, because of severe and moderate impairment of orientation (blind or visually impaired).

---

[1] Hereinafter disabilities are named according to the ICF [7]

E) Needs partial outside care (assistance, supervision), incl. when outside, because of moderate difficulty in self-care and domestic life.

H) Needs partial outside care (assistance, supervision), incl. when outside and guidance by a person who implements care, because of moderate impairment of orientation and/or appropriate behavior.

K) Needs partial outside care (assistance, supervision) and guidance by a personal care provident incl. when outside, because of severe impairment of orientation, difficulties in communication and interpersonal interaction (hearing, speech and vision impaired).

M) Needs professional assistance (sign language interpreter) in formal relationships (mainly outside) because of severe and moderate difficulties in communication and interpersonal interaction (hearing and speech impaired)

O) Does not need any outside assistance. In exceptional cases needs assistance from strangers (personal care providers) experiencing certain life situations when outside.

The second, the third and the fourth levels of the Codifier (body functions + impairment qualifier; body structures + impairment qualifier, activity and participation + appropriate qualifiers) were defined in strict accordance with the ICF [7] (Fig. 1).

The fifth level of description of Technical aids for Disabled persons complied with the ISO 9999:2002 [8] (Fig. 2). A new classification model was developed to define rehabilitation measures and services, which is based on regulations and united the terms used in education, social security, employment, medicine etc.

The structure of the Codifier is a number (chain) of codes, made in accordance with afore listed levels:

$$x/bxxxxx.x/sxxxxx.x/ \ dxxx.x/xx \ xx \ xx1/.../xx \ xx \ xxn1/xx.xx.xx.xx1 /...$$
$$/xx.xx.xx.xxn2 \ ,$$

where x is a code of disability category;
bxxxxx.x is a code of body function with ICF qualifier;
sxxxxx.x is a code of body structure with ICF qualifier;
dxxx.x is a code of activity and participation with ICF qualifier;
xx.xx.xx are the codes of technical aids for rehabilitation defined by the National Standard Specification of the Russian Federation GOST R 51079-2006 (ISO 9999:2002) "Technical Aids for Persons with Disabilities. Classification" [8];
xx.xx.xx.xx are the codes for rehabilitation services defined by a respective component;
n1 is a number of technical aids for rehabilitation corresponding the chosen body function with a qualifier, body structure with a qualifier, disability category, type of activity and participation;
n2 is the number of services corresponding the chosen body function with a qualifier, body structure with a qualifier, disability category, type of activity and participation.

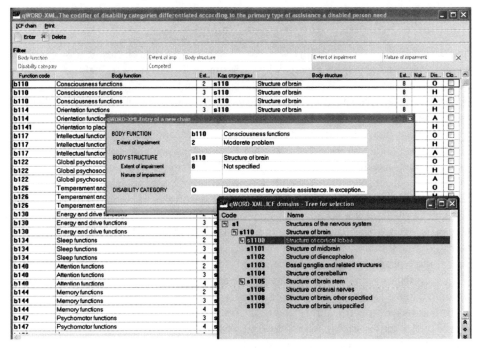

Fig. 1. The example of entering data into the Codifier, the choice of structure according to the ICF

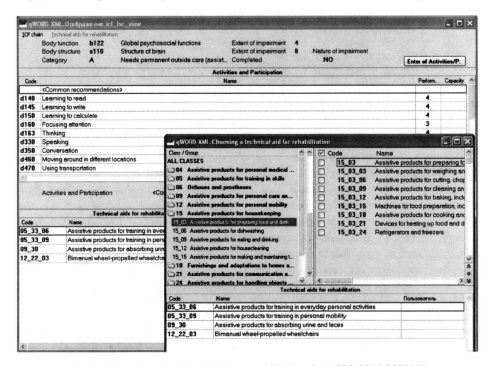

Fig. 2. Choosing a technical aid for rehabilitation from ISO 9999:2002 [8]

4   Information Content of the Codifier

A special computer-aided working place was developed for data entries. It allows to form a chain described above by experts with different skills. Rights were defined for every expert. Some of them had rights to enter new chains and correct the data, access to other experts' data. The head expert looked through all data and had a right for editing them. If the chained was considered as finished, additional editing was restricted.

In the first phase a pilot database was established and all the experts have been trained. The training consisted of two courses: working with the software product and the formation of a different style of thinking. The first course took no more than two days, the second - more than a month. Combination of concepts from the independent classifiers in a single chain required not only a software solution, but also some human efforts. Essentially, new ideology for rehabilitation was proposed: avoiding a diagnosis, objectification by describing the state in terms of body functions impairments, defining the relationship between an impaired function, disability and description of the typical forms of assistance.

In accordance with the ICF a group of experts was singling out body functions and structures related to them which when steadily impaired can lead to disabilities (activity limitations – potential ability and restriction of participation ability – implementation [7]) of a certain extent. The following was defined: a code of disability category, the need of the disabled for technical aids of rehabilitation, rehabilitation activities and services which can help to reduce the disability. Results of the work carried out by experts became the basis for information content of the Codifier.

5   Applying the Codifier

After experts had competed the Codifier a data base was formed using which on the basis of formalized description of body condition according to ICF one can get a list of variants on providing assistance which can be used in making the IRP of a certain disabled person (Fig. 3).

Using the Codifier more than one hundred complex professional terms describing body impairment were managed to be brought to 8, which can be described with common language so that it can be used by those non-specialists in the field of disability. To illustrate that in the given Table 1 there is a number of body functions and structures impairments which require the same type of situational assistance to individuals by the staff of different organizations and companies.

Considering the correlation of most functions, with the help of the Codifier one can fully describe all the impairments of body functions which lead to disabilities. The set of codes of these functions enables to create an individual profile [7] of a disabled person, which can be also a basis for developing of the IRP. After impairments of body functions are defined it is analyzed whether the functions defined by the specialist coincide with those existing in the Codifier, the analysis being carried out at the specialist's computer-aided workplace designed especially for it. In case if the

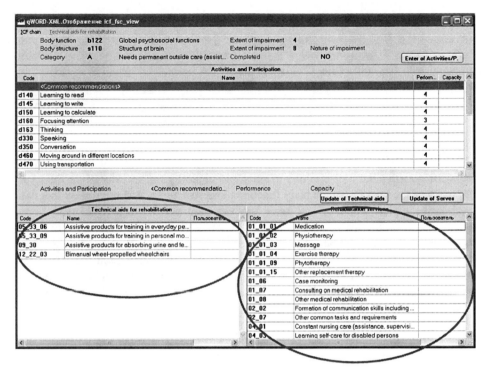

Fig. 3. A list of recommended technical aids for disabled persons (outlined on the left) and rehabilitation services (outlined on the right)

functions coincide, the rehabilitation measures and services recommended by experts of Technical Aids for Disabled Persons are chosen from the Codifier. At the same time notions that coincide are left out. Thus, a specialist get a list of Technical Aids for Disabled Persons, rehabilitation measures and services recommended by experts and based on the Codifier, and which one can follow while working with the disabled. In case if there are no coincidences, a specialist has to choose the necessary

Table 1. Categories of disabilities and disabling functions (ICF)

Letter Code of Disability Category	A	B	C	E	H	K	M	O
Number of Functions' Codes	33	40	2	4	21	3	2	48
Number of Structures' Codes	10	18	3	5	2	2	1	33

measures and services oneself using the respective components that the Codifier includes. Later after the analysis this data may be added to the Codifier.

At a repeated medical-social expertise the evaluation of an individual's profile obtained with the help of the Codifier provides the basis for evaluation of efficiency of the performed rehabilitation measures.

6   Results

The Codifier that we have developed can be widely applied. It is not only an instrument for statistics but it also allows to state and objectify the severity of disability on the basis of the impairment of body functions and structures, and relying on this data to describe typical kinds of assistance that disabled persons need, incl. situational assistance, to define the needed amount of rehabilitation services and technical aids for rehabilitation when making the IRP. Further on, upon evaluation the results of the IRP implementation, relying on comparison of severity of impaired functions and structures, activity and participation of a disabled person before and after carrying out the rehabilitation measures and in activity, one can evaluate the efficiency of rehabilitation of a disabled person.

Thus, with the help of the Codifier the condition of a disabled person can be objectified both from the perspective of impairments of body functions and structures and from the perspective of activity and participation in life situations, evaluate his or her need for rehabilitation, define primary types of assistance to integrate the person into environment. Codes obtained as a result of formalizing the condition of a disabled person can be used to electronically record measures on a social card of a citizen, to control the implementation of rehabilitation measures, to analyze the needs in various kinds of assistance and accordingly to plan a budget for different levels, development of rehabilitation industry etc. This all will raise living standards of disabled persons and improve their integration in the environment.

References

1. Resolution by the Government of the Russian Federation dd. 17 March 2011 No.175 "On the State Program of the Russian Federation "Accessible Environment" for 2011-2015" (Постановление Правительства Российской Федерации от 17 марта 2011 года N 175 "О государственной программе Российской Федерации "Доступная среда" на 2011-2015 годы")
2. The Concept of Improving the State System of Medical-Social Expertise and Rehabilitation of the Disabled (Концепция совершенствования государственной системы медико-социальной экспертизы и реабилитации инвалидов),
   http://www.minzdravsoc.ru/docs/mzsr/handicapped/3
3. Resolution the General Assembly of the UN dd. 13 December 2006 No.61/106 "Convention on the Rights of Persons with Disabilities"
4. Federal Law dd. 3 May 2012 No. 46-FZ "On Ratification of the Convention on the Rights of Persons with Disabilities", (Федеральный закон от 3 мая 2012 года N 46-ФЗ "О ратификации Конвенции о правах инвалидов")

5. InterSystems Corporation,
   http://www.intersystems.com/cache/index.html
6. Dolzhenkov, A., Timofeev, D.: Semantic Tool of Database Development. Open Systems Journal, 1, 39-44 (2006) (Долженков А., Тимофеев Д. Семантический инструмент построения баз данных. Открытые системы, 1, 39-44 (2006))
7. The International Classification of Functioning, Disability and Health. World Health Organization, Geneva (2001)
8. The National Standard Specification of the Russian Federation GOST R 51079-2006 (ISO 9999:2002) "Technical Aids for Persons with Disabilities. Classification" (Национальный стандарт Российской Федерации ГОСТ Р 51079-2006 (ИСО 9999:2002) "Технические средства реабилитации людей с ограничениями жизнедеятельности. Классификация")

Improving Perfect Electronic Health Records and Integrated Health Information in China: A Case on Disease Management of Diabetes

Ren Ran[1,*], Chi Zhao[2], XiaoGuang Xu[1], and Guiqing Yao[3]

[1] School of Public Health, Dalian Medical University, Dalian, P.R. China
renran99@163.com
[2] Dalian Second Hospital, Dalian, P.R. China
[3] University of Southampton, Southampton, UK
G.Yao@soton.ac.uk

Abstract. To examine the role of EHRs in disease management diabetes and other favorable factors, explore feasible strategy for improving the developing EHRs by perfecting top-designing and integration.

The study adopt literature review method, reviews the research literature on EHRs systems. Data of three provinces come from the China Fourth National Health Survey 2008.

Data analysis reveals the gap in diabetes management and the limitation use of data for intervention. Diabetes management has been incorporated into the national package of basic public health service, but not high proportion of EHRs for diabetic patients. It suggest that the effort for establishing links between use of EHRs and disease management is needed for moving forward the integration, due to EHR play unique role for evidence-based on chronic disease management. To accelerate making progress, it not only need to build the implement framework, but the top-designing framework and targeted guidelines.

EHRs' standardization and quality is a key to form good practice. Developing of EHRs' is beneficial chronic disease management and intervention, but need to formulate right policy and practice in perfecting EHRs' system itself and integration with disease management system current.

Keywords: EHRs, Diabetes, Disease Management, Integrated Information, China.

1 Introduction

The Electronic Health Records (EHRs) and the Personal Health Record (PHR) are two innovative health information recording system. EHRs can be used systematically recording comprehensive individual based health information including individual patient health history, social economic background, medication, hospital visits and

---

* Corresponding author.

G. Huang et al. (Eds.): HIS 2013, LNCS 7798, pp. 232–243, 2013.
© Springer-Verlag Berlin Heidelberg 2013

range other medical history over times. Those records can be combined with other routinely collected database such as patients' medical and health care records, family health, public health information, and history of management on chronic diseases and so on. PHR is electronic health information system which records an individual health information and is maintained by an individual patient. Individuals own and manage the information in the PHR, which can be provided from healthcare providers and individuals [1]. Electronic personal health record systems (PHRs) support patient centered healthcare by making medical records and other relevant information accessible to patients, thus assisting patients in health self-management [2]. Such information can be very useful in helping individual patients and health providers in making decision about best treatments. EHRs is created and maintained by the health care provider for the patient. The PHR is created and maintained by the patient/individual for his or her own use.

The EHRs and PHR not only act as effective ways for collecting individual health information, but also realizing the paper health archives "digitization" which will be a core tool in health information management. EHRs have long been regarded as offering significant potential to improve the health system in the United States [3]. Health care industry in general lags behind other sections in adopting information technology by as much as 10–15 years internationally. The usage of EHRs and PHRs are only emerging in recent years. It is evidenced that widespread adoption and meaningful use of health information technology (HIT) can improve the quality, safety, and efficiency in healthcare [4]. HIT can be linked with professional information provided over internet which is proved to foster patient-focused care, and to promote transparency in prices and performance [5]. Hence promote wider adoption of EHR. The first EHR were designed and deployed at late 1960s and early 1970s, following the introduction of the personal computer. By the middle of 1970s, approximately 90% of hospitals used computers in US [6], and estimated that by 2020, approximately 50% of health care practitioners will be adopt EHR.

It is a growing trend in many countries in adopting and advancing EHR. However their use in China is still at its earlier stage. China began to adopt EHR systems and deployments of EHR in recently year and started to use mainly in community health center, as constituent part of .community health service, and also explore HER information sharing and integrated with management of chronic diseases in pilot areas. But China will enter a new era in extending and use of HER along with process of deepening the health care system reform. According to the policy of "Implementation Plan for the Recent Priorities of the Health Care System Reform (2009-2011)", promoting the gradual equalization of basic public health services, the items of basic public health services will be defined and the content of services specified, including establishing health records.

2   Data and Study Methods

The study adopt literature review method, reviews the research literature on EHR systems, including review literatures in international and national level. The Literature

234   R. Ren et al.

research examined the development and use of EHRs, and role an advantages of EHRs, and integrated with disease management, as well as the interaction of EHRs emerging for health system and health reform. A case study of diabetes management using EHRs was discussed in the paper.

A field survey carried out in four community health center in Liaoning Province, by using observation, data collection and interview survey in township hospitals in rural area and community health center in urban area. Interviewer consisted of faculty and students of Medical University and training before the survey. The data collection and observation focus on health records system and chronic disease management, such as total numbers of health records being fill in, and therein for diabetes patients, numbers of diabetes patients, and numbers of team undertaken EHRs and so on. Interviewing survey aim at investigated the center' general information on health records system and issues or obstacles for developing EHRs.

Comparison analysis on established electronic health records in three provinces used for measuring the progress and differences among different area, including average level for diabetes management and implementation health information function in township hospitals. Data come from the China Fourth National Health Survey 2008.

3   Results

3.1   Advantages of EHRs and Development in US and Other Countries

Among the health information systems, EHR has long been recognized the main source of health information. Recent years as an important tool of chronic disease management, the function and value of EHR become even more prominent. One of most prominent advantages of EHR is beneficial to the individuals self-management health and identifying risk factors. Chronic disease self-management programs seek to empower patients both by providing information and by teaching skills to improve self-care and provider-patient interactions. The EHR can assist patients in managing their health condition through individualized care plans, graphing of symptoms, passive biofeedback, tailored instructive or motivational feedback, decisional aids, and reminders [7].

Ongoing health care improvement and patient safety initiatives demand new information collection and communication technologies. Pressing needs of cost-effectiveness in healthcare and opportunities of emerging electronic health record technologies offer unprecedented chance for progress. It has been shown that EHRs not only improve the quality of patient care, but also provide important information for health policy planning [8].

As a useful tool for managing relevant health information, it can play important role for promotion health maintenance and assisting with chronic disease management via an interactive, common data set of electronic health information and e-health tools. In several studies has been shown that the use of an information system was conducive to more complete and accurate documentation by health care professionals. When patient data and medical knowledge are accessible electronically, decision support technology can improve all types of health care decisions and transform health care [6]. Especially along with the increasing of the prevalence rate of chronic diseases, disease managements, implementation and monitoring intervention need to health information. The NHS information strategy, the national service frameworks and the NHS plan all promote the use of electronic patient records. The NHS national specification for integrated care records service aims to develop clinical records, which are to be designed around the patient, integrated across all health and social care settings, and capable of supporting the implementation of care pathways within the national service frameworks [9].

Recognizing these usages, many countries and the World Health Organization have launched several major health care improvement initiatives that are driven by new electronic record technologies. $19 billion investment provided for it in the United States in 2009 [10]. President Bush announced the goal of assuring that most Americans have EHRs within the next 10 years on April 26, 2004. In a lot of countries, such as the United States, Australia and the UK, the purer EHR model is evolving at the national level. In addition to projects of national scope, some state governments of US have EHR launch initiatives; for example, Massachusetts has recently announced a statewide initiative, with the goal of having a statewide electronic records system in place within five years [10]. Australia's Health Connect respects patient and provider choices and generates only limited data sets, the US system are moving towards interoperability and comparability of all patient data, maximizing patient data flow into local and national systems.

3.2 Development and Goal of EHRs and Health Informatization Construction in China

The beginning of Health records started early 80's of last century in China as the development community health service. It developed rapidly nationally after it become one of content among the community health services under the policy document "The Resolution of Health Reform and Development" in 1997 issued by CPC and State Council, especially in eastern area, such as Shanghai, ZheJiang province and so on, add the behavior factors of chronic disease (hypertension, diabetes and so on) into community's health records. Shanghai designed and popularized the software for health records in 2000 [11].

There have two drives for pushing forward developing EHR in China. The first drive is health informatization construction strategy, and second is pushing policy for equalization of basic public health services. The health informatization construction has been placed on the national development plan and reform schedule in China, in addition to EHR as a embody of health informatization and constituent, infrastructure construction and technique advances of building health informatization provide feasible basis for developing EHR. Promoting the gradual equalization of basic public health services, a significant target of Chinese health reform, contain health records as basic items for the package of public health services. This starting from 2009 residents' health record will be gradually established with standardized management nationwide. A national project of basic public health service has been launched since 2009, after that establishing health records widely carried out in grassroots health institutions both in urban and rural. Among eleven of public health services in the document in 2011 Edition, health records management for urban and rural residents and the establishment of individual health records have covered in the package. Moreover, the policy stressed the chronic disease information recording (such as diabetes) involving health records management.

Particularly there have formulated several related policies in moving forward health records system. After formulated document of "Specification for Drafting of Basic Dataset of Health Records" in 2009 by MoH, the goal in twelfth five-year development plan for health information proposed by Chinese central government: and proposed "by 2015, the framework of health information system will be build, the progressive realization of a unified and efficient interconnection, gradually establish a shared health records and electronic medical record database, for 30% of the residents establish health card and conform to the standard of electronic health records". In the document, it also emphasized to build individual electronic health records database as one of key tasks for Health informatization construction. Minster of Health, Chen Zhu proposed in the Second Health System Research Symposium this month, "Speed up Health Informatization"; Establishing practical and shareable health information system, promote interconnection and resource sharing; Promote the application of resident electronic health record and medical record: promote database of electronic health record and hospital information system.

There carried out the EHRs usage research for community setting in Pu Dong district. It has selected 435 data for new EHRs among 601 data, including basic information on chronic disease .management, such as diabetes' information (diagnosis, blood glucose and so on). The most prominent progress in there is realizing the information sharing by establish information platform among pilot health agencies (3 hospitals, 3 community health center). Providers in the system can share patients' HER information by on line way, such as look up the patients visits numbers and related information [12].

3.3   Role of EHRs in Diabetes Disease Management and Integrated of Health Information

EHRs were used in chronic disease management of primary, secondary and tertiary care. A typical case is diabetes mellitus. A lot of researches have showed that EHR can improve the quality of care and patient outcomes, and contribute to the health of the population, in terms of effective intervention measures after collecting the diabetes patients [13-15]. Researchers suggested that it is effective to intervention diabetes and other chronic diseases in community, and the United States, Canada, South Africa, Israel and other countries achieved performance in the area since the late twentieth Century [17]. EHR models present significant advantages because of their potential to deliver a longitudinal record that tracks all medical interactions by a particular patient and provide comprehensive data across populations.

The disease management of and health information recording of diabetes in China is still in the exploratory stage, despite governments encouraged that health records of chronic diseases in community implemented cover all of diabetes patients, and set up the financial incentives for collecting and filling health records in the national policy. Due to lack of standardized guideline on management and implementation procedure, as well as personnel capacity both the quantity and quality of the task are affected.

As a basis for screening diabetes in community, EHR is an important tool, particularly useful in identifying timely the high risk diabetic groups and screening undiagnosed diabetes. Research has indicated that there are about 10 years before clinical diagnosis of diabetes, and diabetic patients with undiagnosed exist large populations, and estimated about 50% in western countries [18] and 70% reported by Shanghai data [15]. Obviously patients with undiagnosed diabetes were not able to obtain timely intervention through the dynamic health record information and supplemented by the corresponding inspection, can be effectively detected those high risk population, providing chance for adopt effective prevention measures as soon. Also EHRs is an important information source of diabetes risk evaluation and grading of the disease management, and improving the quality of diabetes interventions and the management process. The community health center will make dynamic disease management of diabetes patients based on the blood glucose level, risk factor, target organ damage situation from the HER's providing information, as a basis for classification management of risk factors in patients with diabetes, and make it possible for timely monitoring of interventions.

Furthermore, it can integrated of related information and data to different medical information among different agencies, and realize by exchange and sharing the update and interactive health information; so as to improve the information application in the disease management and providing integrated care. In replying to application, recently US have focused primarily on the technical aspects of EHR implementation. Considerable development is underway to standardize event taxonomy express knowledge representation such as clinical practice guidelines [10].

Besides, emerging electronic health record models present numerous challenges to health care systems. It promote a new revolution for redesigning the model of chronic disease (like diabetes) management. A new trend of EHRs is to be catalysts for redesign care and informing changes required to ensure quality health care in the new century [19].As the Institute of Medicine's recommendations for care redesign, tenets of care redesign included providing tailored care for patients, recognizing the patient as the source of control of care, sharing knowledge, decision making based on current evidence, transparency in care, and clinician cooperation. The era of health information technology has developed rapidly and continues to advance, and technological solutions have been implemented to enhance consumer access to EHRs and PHRs [20].As one that provides access to all or most of the patient's clinical information, both EHRs and PHRs support patient centered healthcare by making medical records and other relevant information accessible to patients, thus assisting patients in health self-management.

3.4   Problems and Gaps in Development and Integration of EHRs

Although China has promulgated a series of documents on health informatization construction (HIC) after health reform initiates, and HIC has become one of main indicator in health reform, its implementation in disease management and health promotion remain in theories. The targets of establishing HIC has been put forward, but in general lack of effective actions and recommendation on how to in adopt EHR in chronic diseases management. Unidentified technical problems are common in implementation procedures for using EHR, such as lack of technical guidelines for usage of EHRs and integrated EHR in chronic diseases management, and path of advancing HER in HIC implementation.

Along with increased prevalence and incidence of chronic diseases such as diabetes, chronic diseases management have become a major problem facing health care provider and affecting health of Chinese population in China. The rate of morbidity and mortality continues growth, Along with the rapid change in lifestyle and the rise of income level in China, diabetes has become a high incidence. The prevalence of total diabetes (which included both previously diagnosed diabetes and previously undiagnosed diabetes) and pre-diabetes were 9.7% and 15.5%, respectively, accounting for 92.4 million adults with diabetes and 148.2 million adults with pre-diabetes [21]. To identify risk factor and provide preventive intervention could delay the onset of such disease, it is crucial for "early discovery", "early treatment" and "early intervention", based on individual risk factors records and adopts effective intervention measures. EHRs can make it possible by recording the personal health information, identify risk factors and monitor blood glucose changes, as well as tracing intervention implementation progress and evaluation pre-diabetes and diabetes patient management.

Nevertheless diabetes management at present has been incorporated into the national package of basic public health service and the government for the provision of grants, there are not high proportion of EHRs for diabetic patients in the community in our survey, most of them are not dynamic and update regularly. Moreover, we found that relatively little diabetes patients have been really tracked by community workers using health records, and those registration patients just as "the tip of the iceberg". Also, even for those groups being recording information have been not yet effectively intervention in some area. The data from Research on Health Services of Primary Health Care Facilities in China [22] showed, system of health records have not associated with the diabetes disease management, diabetes disease management rate only 28.7%, and the highest is area of eastern with 47% and the lowest is western, with 19.4%. Further, the information, including the risk factors of diabetes and personal health records, collected by communities has not been integrated with hospital's medical records systems. When diabetic patients accepted to higher levels of the hospital for medical treatment, these EHRs are not integrated into the Electronic medical records in higher level hospitals to provide reference for a doctor in diagnosis and treatment.

Yet currently filing rate of the electronic health records was very high in China, in Shanghai, a district of Minhang District as an example, this only includes the permanent residents 826700 with coverage rate of 90.31% in 2008. However it does not provide any information related to diabetes management. In Shanghai there are 116000 diabetes patients with blood glucose control rate at 21.63% [23].

A National Survey on Health Services of Primary Health Care Facilities in China in 2008 shows that progress on establishing health record and diabetes management in township hospitals. There are difference among three area, and the most high percent of establishing health records is eastern area (46.2%), the lowest is Western area (18.7%);

screening and disease management of diabetes have similar situations (Fig.1). Table 1 also reflect the difference for implementation health information function in township hospital between three provinces (Table. 1), such as "network & electronic filling for health information", Shandong and Hubei with more low coverage than Chongqin. Furthermore, the value of data are not only reveal the big gap in diabetes management among province, but also display the limitation use of data for decision making and intervention.

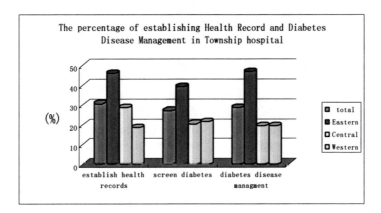

Fig. 1. The percentage of establishing health record and diabetes disease management in township hospitals

Table 1. The Percent of Implementation Health Information Function in Township Hospital (%)

	Total	Shandong	Hubei	Chongqing
network & electronic filling for health information	31.4	20	20	60
based on data formulating plan of health promotion and disease intervention	31.4	26.7	10	60
set up community's health information system	25.7	20	10	50

The main problems of health information in diabetes management in community implemented of China at present include: (1) lack of integration system of diabetes information, not yet established effective of the electronic information network among community-CDC-hospital; (2) lack of complete, continuity information, and not to carry out personal health information collection and update based on EHRs, and follow-up in community. For example, community health workers need to 4 times a year follow-up of diabetes mellitus patients according to the requirement benefit package of national basic public health, but the follow-up information did not integrated into diabetes patient's EHRs even it have done. In fact most of health records in there always keep data or information from beginning to end. (3) EHRs has not been associated with diabetes disease management and integration care of higher level, even EHR is established for diabetes patients, did not fully play the role of EHR in diabetes's intervention and disease management.

EHR universal coverage and sufficient information on personal health and risk factor are premise for promoting the management or intervention effective and timely. Only a scientific, systemic and integrated health information system on diabetes can fully improve the intervention and disease management. Therefore it is necessary to improve designing of health information system and EHRs standardization so as to achieve the national goal.

4   Challenges and Difficulties for Perfecting EHRs and Integration of Health Information

There are several challenges for further building and perfecting EHRs system and integrated information in China, so the effort should be devoted to what could be the most promising strategies and approaches to the meeting the challenges, and set priority area by wisdom decision making . Firstly, strengthen top-level design and standard development of health information system, not only put forward the standardization requirements, but choose some good model as reference. Second, integrate existing information system across facilities and data of EHRs in community and medical records in hospital. Third, priority the key groups for establishing EHRs system and integrated information, some chronic disease patients, such as diabetes, hypertension, being the preferred option. China government focused efforts on building strategy of national health information technology ten years ago, and issued document of the Framework of National Development Planning for The Health informatization (2003-2010) in 2003 by MOH, and proposed the principles of building the health information system with perfect function, standardized norm, and credible safety.

The core value of EHRs is to use the therein information. The fundamental goal and key of health informatization construction just application information effectively, and build a platform of integrated all of health information and risk factors according to the needs of disease prevention and treatment, and intervention monitoring. Our preliminary results, along with other related research, suggest that the effort for establishing links between use of EHRs information and disease management is needed, both to determine how to benefits from critical health information use and to identify what mechanisms for pushing EHRs use and building the links, in keeping with national policy and target. EHR plays unique role for evidence-based on chronic disease management. To accelerate making progress, it not only need to be EHR as an information management tools, but more need to build the implement framework for connecting EHR and chronic disease management by constructing, establishing the integration mechanisms.

The difficulties use of EHR in chronic disease management is how to realize the tracking and sharing the health information with standardized, orderly, updated, achieving the interaction of different institutions, integration of health information on different medical institutions and different health services, e.g., outpatient and hospital, community health service centers and hospitals, CDC and community health service centers [24].

For the strategies on prevention and control of chronic disease management by forward lead direction, it needed to implement the system management and intervention in the community as the center for prevention interventions of risk factors and systemic management of chronic disease, by means of the EHR to track the progress and data with normative, orderly and continuous monitoring. A lot of reviews point out trends that show stronger involvement of the patients-citizens in the health care prevention and promotion processes [14], and the future development of the electronic healthcare record into personal health records.

It is imminent not only to advance EHRs of diabetic patients and information system at present, expending the EHR system to all of diabetes patient, but promote the quality of health information collected and by pushing EHRs standardization, such as formulating guideline and rules, so as to provide the basis for improving the quality of EHRs information and interventions. It is need to concern the integration of a full, all-round health information, gradually realizing information interconnection and interworking for all of chronic disease, as well as data interoperability across health institutions, fully realizing information sharing for interagency and trans-regional, in order to improve the disease management and implementation of interventions more cost effective and good performance.

To accelerate progress in this area, we believe that more strategies for meaningful use of EHRs should be considered. The strategies meet the challenges are as follow: (1) Formulate standardized rules and guiding principle for promotion good practice of EHRs. Formulating standardized rules and guiding principle is a prerequisite for ensuring the quality and consistency in developing function and expanding usage of EHRs. The use of EHRs and good practice based on sound standardized system and clear implementation rules, and set up suitable incentives, include financial and nonfinancial incentives. (2) Distinguish different target groups of diabetes disease management, and establishing the electronic health record, such as high risk groups, diagnosed patients with diabetes mellitus and the diabetes patients with complications, and follow-up and regularly updated content of EHRs. (3) Integrate into EHR and follow-up information of diabetes patient, through the electronic information technology to achieve the information integration. (4) Realize Information sharing of diabetic disease management among EHR in community and CDC as well as medical records of hospitals, shaping a bidirectional interaction network of diabetes disease management. Based on the above information, establish a interconnect system of computerized, dynamic management, regularly updated content of health care records, submitted to the integrated diabetes information network.

5   Conclusion

The use and drives of EHRs are likely to increase markedly by 2020 in China, as moving forward the package of basic public health service and health informatization construction in health care reform. One of key usage of EHR is to management of chronic diseases. EHRs standardization and quality is a key to form good practice. The use of EHRs information and other information sharing initiatives depend on the top-designing framework and targeted guidelines. Only EHR with good quality and targeted perfect designing can reach the good practice and play right role in health system as well achieving the goal of health reform.

It is anticipated that the developing of EHRs will have major beneficial chronic disease management and intervention, but depend on current right policy and right practice in perfecting EHRs system itself and integration with disease management system.

It is urgent to formulate policy and guidelines for perfecting the top-designing framework. The pressing challenges are how to promote the EHRs quality and make the best of usage, as well as perfect top-designing framework and realizing information sharing and interactivity. Also, the technical aspects of EHR implementation need to pay more attention at present, such as identifies structures and standardization of EHR system, define the function and integrated path with chronic disease management, and priority strategies. A further challenge is needs and requirements of different health care professionals and consumers in the development of EHRs, and the use of terminologies in order to achieve interoperability.

References

1. AHIMA eHIM Personal Health Record Work Group. The role of the Personal Health Record in the HER. Journal of AHIMA 76(7), 64 A-D (2005)
2. Archer, N., Fevrier-Thomas, U., Lokker, C., McKibbon, K.A., Straus, S.E.: Personal health records: a scoping review. J. Am. Med. Inform. Assoc. 18(4), 515–522 (2011)
3. David, F., Lobach,Don, E.: Detmer, Research Challenges for Electronic Health Records. Am. J. Prev. Med. 32(5), S104–S111 (2007)
4. Mery, B., et al.: Anticipating and addressing the unintended consequences of health IT and policy: a report from the AMIA 2009 Health Policy Meeting. J. Am. Med. Inform. Assoc. 18(1), 82–90 (2011)
5. Goldschmidt, P.G.: HIT and MIS: Implications of Health Information Technology and Medical Information Systems. Communications of the ACM 10, 9–10 (2005)
6. Institute of Medicine., The computer-based patient record: an essential technology for health care. In: Dick, R.S., Steen, E.B., Detmer, D.E. (eds.), 2nd edn. National Academy Press, Washington DC (1997)
7. Jeremiah, G.R.: How the Electronic Health Record Will Change the Future of Health Care. Yale J. Biol. Med. 85(3), 379–386 (2012)
8. Häyrinen, K., Saranto, K., Nykänen, P.: Definition, structure, content, use and impacts of electronic health records: a review of the research literature. Int. J. Med. Inform. 77(5), 291–304 (2008)
9. Julia, H.-C., et al.: The electronic patient record in primary care—regression or progression? A cross sectional study. BMJ 326, 1439–1443 (2003)
10. Balas, A., Al Sanousi, A.: Interoperable electronic patient records for health care improvement. Stud. Health Technol. Inform. 150, 19–23 (2009)
11. Gunter, T.D., Terry, N.P.: The Emergence of National Electronic Health Record Architectures in the United States and Australia: Models, Costs, and Questions. J. Med. Internet. Res. 7(1), e3 (2005)
12. Wu, W.D., et al.: The Development and Study Trends for Community Health Records in China. Chinese Health Statistics 24(8), 444–446 (2007) (in Chinese)

13. Du, Z.H., Peng, H.Z.: Research of Application of Community Electronic Health Records in Pudong New Area. Chinese J. of General Practice 8(2), 207–208 (2010)
14. Iakovidis, I.: Towards personal health record: current situation, obstacles and trends in implementation of electronic healthcare record in Europe. International journal of medical informatics 52(1-3), 105–115 (1998)
15. Li, R., Li, X.J., Wang, Z.G.: The Feasibility for Using Health Records in Screening and Management of Diabetes in Community. J. of PHM 18(3), 235–237 (2002) (in Chinese)
16. Huang, X., Fan, Q.Y.: The Role for Realization Chronic Disease Objective Management of Electronic Health Records. Chin J. of General Practice 6(8), 863–864 (2008) (in Chinese)
17. Zhang, S.J., et al.: The Preliminary Study of Type 2 Diabetes on Community Comprehensive Prevention and Cure. Chinese General Practice 8(15), 1256–1257 (2005) (in Chinese)
18. Engelgau, M.M., Thonpson, T.J., Aubert, R.E., et al.: Screening for NIDDM in nonpregnant adults. Diabetes Care 18, 1606–1608 (1995)
19. Institute of Medicine, Crossing the quality chasm: a new health system for the 21st century. National Academy Press, Washington, DC (2001)
20. Caligtan, C.A., Dykes, P.C.: Electronic Health Records and Personal Health Records. Seminars in Oncology Nursing 27(3), 218–228 (2011)
21. Yang, W.Y., et al.: Prevalence of Diabetes among Men and Women in China. N. Eng. J. Med. 362(12), 1090–1091 (2010)
22. Center for Health Statistics and Information of MOH, Research on Health Services of Primary Health Care Facilities in China. Peking Union Medical University Publishing, Beijing (2008)
23. Lu, J.G.: Development of Regional Health Information Centred Electronic Health Records, http://wenku.baidu.com/view/3f13fa1a964bcf84b9d57b65.html.2010-1
24. Zhong, N., Wang, H.Q., Chen, D.D.: The Development and Interactive Applications in Electronic Medical Records and Electronic Health Records. Chinese Journal of General Practice 8(10), 1318–1319 (2010) (in Chinese)

Payment Information Calculation System (PICS) Healthcare Quality Measure Specification

Denise Magnani and Michael Freed

General Dynamics Information Technology, Health Analytics and Fraud Prevention,
West Des Moines, IA and London, England
© Vangent, Inc., a General Dynamics Company 2012
{denise.magnani,michael.freed}@gdit.com

Abstract. Measurement of healthcare quality in multiple care settings is a top priority. The ability to keep pace with the evolving standards of care to ensure we are measuring the right things at the right time is critical. The flexibility to add, update, and retire quality indicators easily and in tight timescales is essential. The Payment Information Calculation System (PICS), utilized in multiple national contracts for healthcare quality data collection and calculation, is facilitating the ability to respond. The design of the system enables increased efficiency by incorporating rules-driven technology. The tool provides non-developer subject matter experts the ability to define clinical data collection business rules in an XML format used by the PICS consumer to generate data collection user interfaces. This flexibility provides the ability to construct new data collection modules complete with data capture definitions, flow-logic (skip patterns) and enforce data collection formats and validate captured data (edits).

Keywords: healthcare quality, measurement, technology, data collection, flexibility.

1 Introduction

Measurement of healthcare quality is a top priority. The ability to keep pace with the evolving standards of care to ensure we are measuring the right things at the right time is critical. The flexibility to add, modify, and retire quality measures/indicators easily and in tight timescales is mandatory. To meet these needs, we have created the Payment Information Calculation System (PICS) that can define measure specifications across any setting of care for rapid implementation. The PICS tool is facilitating the ability to respond as desired by the stakeholders. It is a proven tool that has significantly reduced the Software Development Life Cycle.

2 PICS Framework

The PICS framework was developed to meet the continuing need to rapidly build and deploy applications responsible for the remote capture of patient and healthcare

G. Huang et al. (Eds.): HIS 2013, LNCS 7798, pp. 244–249, 2013.

provider data directly from clinical settings. PICS meets this need by providing the core components required for this type of activity such as security, user management, patient management, database, reporting and healthcare provider (facility) components.

In the past, the healthcare quality indicator specifications were hard-coded into each of the various data collection and database tools. This did not allow for flexibility, the ability to react quickly to changes in healthcare standards, or consistency amongst the products.

To resolve this issue, PICS contains a Module Developer component providing non-developer subject matter experts the ability to define clinical data collection business rules in a PICS specific XML format used by the PICS consumer to generate data collection user interfaces. This flexibility provides the ability to construct new data collection modules complete with data capture definitions, flow-logic (skip patterns) and enforce data collection formats and validate captured data (edits). Since nearly all of the business rules of this data capture are contained within the PICS generated XML, updates and changes to quality measure specifications are easy to perform and manage. The definition of the specifications are then consumed by other system components (i.e., a clinical warehouse), which allows for a single one-stop shop for specification definition to ensure consistency throughout the system.

The PICS core application provides the java based database, application and presentation layer in a three-tier package that runs self-contained on a user's desktop computer. The application is highly compliant with architecture/design and security standards. Use of PICS provides the multiple quality initiatives with a significant head start on compliance, security, database design, reporting, user management, demographic (patients and organizations), import and export abilities and advanced business rule definition and processing.

3   PICS User Interface

The PICS user interface, which non-developer subject matter experts utilize to define quality measure specifications, allows for the definition of initiatives, topics, indicators/measures within each topic, and the associated details to include the data elements needed to calculate the indicator/measure, any associated skip logic (i.e., if the patient is allergic to a medication, it would not be given), edit logic, and the detailed calculations to determine the performance outcomes. The PICS user interface also allows for the preview and test of the indicator/measure definition and how a measure calculation appears and performs. There are numerous calculation methods defined within the tool that allow for easy selection and application. These vary from numerator/denomination percentage calculations to extensive prevalence and adjustment methods.

3.1   Defining Measure Specifications

The PICS Measure Set Editor allows the user to create new measure sets as well as update previously defined measure sets. The user may group measure sets under a higher domain level (i.e., clinical versus patient experience) and roll up the domain levels into an overall quality initiative. The user may define multiple versions of the initiatives for different time periods or financial years.

The Measure Set Editor displays the overall modules contained within the designer in the left pane and all of the indicator groups defined in the right pane. The user may drill down into each indicator group to view or update the individual indicators (See Fig. 1. Below).

Fig. 1. PICS Measure Set Editor

Once a user has selected a measure/indicator, they can view the item details (see Fig. 2 below). This displays the question code, question text, short question title, answer type, initially enabled, if a look-up table is utilized (i.e., ICD Codes, Medications, etc.), if the indicator is required, and if it is to be exported in the XML file. The appropriate answer values can also be defined for each of the data elements needed to calculate the measure.

At any time throughout the measure creation or updating process, the user may preview the data collection elements by selecting the Preview tab from the Measure Set Indicator. This will display how the data collection has been defined in a user-friendly format.

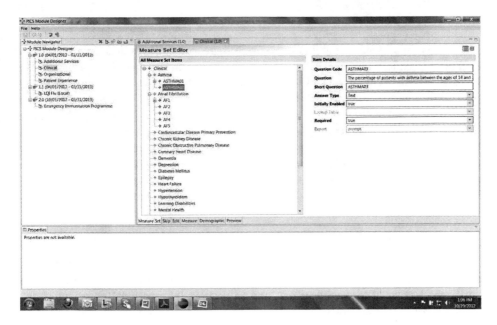

Fig. 2. Item Details for a Specific Indicator (ASTHMA03)

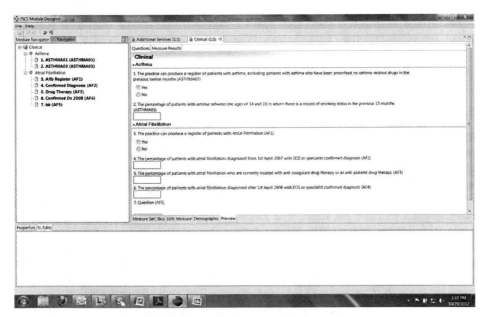

Fig. 3. Measure Data Collection Preview

3.2   Defining Skip Logic

The PICS tool provides the ability to add skip logic to the data collection questions to enable or disable a question based on the answer to another question. A basic skip logic statement includes a Question, a Decision, and a Result. The Question is the data collection question that will be the starting point to build your logic from. The Decision is a statement in which your skip logic, or condition, is written. The Result defines what other question(s) you are enabling, or disabling, based on the result of the Decision. For example, if an indicator is measuring if a certain medication was provided within an identified time frame, if the user answers that the medication is contraindicated, the remaining questions may not be valid. Applying skip logic allows the user to enable or disable data elements as the facility provides responses.

3.3   Defining Edits

The PICS tool also provides the ability to add edit logic to the data collection questions to prevent the user from entering data that does not logically make sense, to verify the format of answered values, and to provide informational messages if desired. A basic edit statement includes a Question, a Decision, and a Result. The Question is the data collection question that will be the starting point to build your logic from. The Decision is a statement in which your edit logic, or condition, is built. The Result defines what edit message is displayed to the user, based on the result of the Decision. The edits can be user-defined, and/or reference look-up tables for validation (i.e., ICD Codes). The user may assign a severity to each edit – whether it is a critical edit that requires user intervention or is an informational edit that may warrant user intervention.

3.4   Creating Measure Calculations

The PICS tool provides the ability to add measure calculations/outcomes for individual measures, and/or groups of measures. In the PICS tool, measure calculations/outcomes are built from the analytic flowcharts contained in the measure specifications or business rules. A basic measure statement includes a Measure, a Decision, and a Result. The Measure is the individual measure for which will be the starting point to build the logic. The Decision is a statement in which the measure logic, or condition, is built. The Result defines the measure outcome for the measure based on the result of the Decision. PICS will allow users to define calculation methods, including thresholds, numerator and denominator specifications, adjustment methods (i.e., risk adjustment, patient set adjustment), conversion methods, and forecast methods. The measure calculation definition also allows for exclusions and exceptions to be defined that may exclude the data from the denominator.

3.5   Exporting Measure Specifications

Once the measures are defined within PICS, they can be exported in XML format to allow for consumption by other tools/databases. For example, the XML files are used to build the data collection user interface at the facility level and also used at the clinical warehouse to validate and calculate measures upon submission by the facility. This allows for the measure specifications to be defined once and the propagated out to other tools which utilize this information.

4   Value-Added Results

PICS enables non-programmer business users to technically describe each quality service, the associated indicators, data elements, and manual collection rules such as skip and edit logic, if applicable, and the calculation routines. The utilization of PICS has significantly reduced the Software Development Life Cycle. The development cycle for the measure specifications dropped from approximately 3 months to 3 weeks. This allows for measure specifications to be updated in a more timely manner and allows for increased testing ability.

Comparisons of Dynamic ECG Recordings
between Two Groups in China – A Preliminary Study

Yang Li[1,2,3,4], Yan Yan[1,2,3], Leilei Du[1,2,3], Yu Luo[1,2,3], Wencai Du[4], and Lei Wang[1,2,3,*]

[1] Shenzhen Institutes of Advanced Technology, Chinese Academy of Sciences
[2] Shenzhen Key Laboratory for Low-Cost Healthcare
[3] Key Laboratory for Health Informatics of Chinese Academy of Sciences
[4] Hainan University
wang.lei@siat.ac.cn

Abstract. Geographic medicine is becoming important because it aims to provide an understanding of health problems and improve the health of people worldwide based on the various geographic factors influencing them. The research of geographical environment with health status is getting higher practical significance. In this paper we presented a cross-regional comparison experiment for dynamic ECG recordings from two distinct groups: one group is in Northeast China (NEC), the city of Shuangya Hill, Heilongjiang; another group involves subjects in the Shenzhen Institutes of Advanced Technology (SIAT), group subjects were from different place of China. Conventional Heart Rate Variability (HRV) parameters, such as SDNN-24, RMSSD, PNN50, SDANN index, SDNN index (time-domain indexes), TP, ULF, LF, UF, LF/UF (frequency-domain indexes), and VP, VT and AT (arrhythmia indexes) were analyzed. The results show that arrhythmia indexes of the two groups have a significant difference which reflects the health conditions are quite different. Accordingly, with further analysis of the frequency-domain indexes and time-domain indexes, it shows that the time-domain indexes were almost the same while the frequency-domain indexes shows there were a 20% average difference between the two groups. Apparently, for the analysis of HRV, the frequency-domain indexes are more persuasive than time-domain indexes.

Keywords: Geographic medicine, Dynamic ECG.

1   Introduction

Medical geography is an integrative, multi-stranded discipline, which studies the relationship of the distribution regularities of the crowd disease and health condition with geographical environment [1]. With the improving of life quality, people pay much more attention to health statuses; the research on the health status with the geographical environment becomes more important [2-4].

Our team in Chinese Academy of Sciences (CAS) developed the Hai-Yan program: key technological breakthroughs sea terminal and cloud platform which are applicable

---

* Corresponding author.

G. Huang et al. (Eds.): HIS 2013, LNCS 7798, pp. 250–257, 2013.

for basic healthcare institutions, and then set up healthcare industry with Chinese characteristics. Lots of experimental units were set to realize Low-cost healthcare [5]. As an important part of the program, a research on healthcare and medical geography has high theoretical significance and significant using value because of the population distribution and the different geographical conditions. As the leading cause of death, the cardiovascular disease treatment means should be considered carefully, a same treatment method can never adopted to different region sufferers. The clinical trial in this paper researched about the HRV parameters' otherness due to environment difference in two groups: NEC and SIAT, in which the SIAT group was adopted as the control group to analyze the cardiovascular disease characteristics in the NEC group aim to provide a healthcare geography reference for Hai-Yun program.

HRV is considered to be a main index to reflect the autonomous nervous system function state effectively; it is an important mechanism of steady state regulation about the cardiovascular system [6]. The HRV increases as the sympathetic or vagus nerve turns more active. Arrhythmia (cardiac arrhythmia) refers to the heart's impulse frequency, rhythm, origin place, conduction speed or excited order abnormalities [7-9]. It is the important index of angiocardiopathy judgment. Therefore, analysis of HRV and arrhythmia has a very important significance for body health condition judgment. The HRV (Heart rate variation) refers to the time duration between consecutive R waves of electrocardiogram (ECG) [10]. Autonomous nerve dysfunction plays a very important role in all kinds of cardiovascular disease onset, development and prognosis of illness [11, 12].

2   Background

China is a vast country with a large population. In recent years, China has witnessed rapid economic development and great increase in citizen and rural living levels. Healthcare and medical treatment are confronted with great challenges. Since countries consists of different kinds of landform or climate such as US, China and India etc., the treatment and healthcare strategies and polices should be considered separately. The morbidity patterns in developed countries especially in similar-environment countries should not reject the further research in diseases differs from the regional condition.

For the developing countries, nosogeographic problems become much more urgent nowadays, direct and useful guides are needed to assure efficient in execution and low in cost. The Hai-Yun program is a program proposed by Chinese government and Chinese Academic of Sciences which aims at providing a low-cost healthcare which had already set more than 4,000 experimental units in China, researched on the medical and healthful geography is urgent for the efficiency in policy formulate and program layout [5]. Moreover, the clinic trial results analysis provided an application example of conventional analytical method.

Medical geography deals with the relation of climate and environment to physiology and its distribution in space within any given period of time. The aim of geomedicine would be to map out not only the distribution and peculiarities of the disease, but also the status of medicine and sanitation over given areas of space, within the same time-interval.

3   Methods

During this study, a conventional experiment method---using dynamic electrocardiogram monitoring technology were used to acquire electrocardiograms from the subjects, and then the 24 hours dynamic ECG time-domain parameters and frequency-domain parameters were analyzed, with a different religion subjects were adopted. The adopted device for the clinical trials was 12-lead Hotler and the participants included 2 groups: the NEC group and the SIAT group. Firstly, the HRV parameters of the two-group records parameters were analyzed, the results shows that the NEC group has a significant lower performance. Moreover, the frequency-domain indexes and time-domain indexes parameters were analyzed.

3.1   Subjects

The two groups included about 50 subjects: the NEC group included 20 subjects which were from the city of Shuangya Hill, northeast of China with a cold weather; the SIAT group involved 30 subjects from students and employers from our institute.

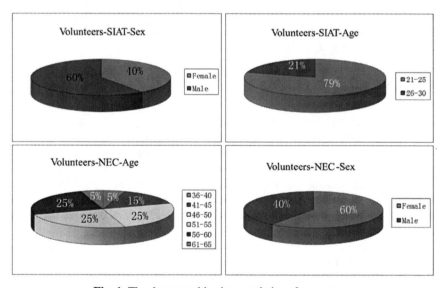

Fig. 1. The demographic characteristics of two groups

As the age and sex information of subjects shows, age range of group SIAT (control group) was 20 to 30 while the experimental group (NEC group) was 30 to 65. The subjects' sex rations were 60% to 40% and 40% to 60%. Moreover, the SIAT group subjects were from different places of China and the NEC group subjects were all from northeast China. The following figure is the map of China, as it shows, the distribution of the subjects included the SIAT group which represented by round spots means SIAT group subjects were from different place of China and the NEC group used five-point star means that NEC group subjects' regions were all in the Northeast of China (the city of Shuangya Hill, Heilongjiang Province). The population distribution of the control group was logical to explore the geomedicine characteristics of the experimental group.

Fig. 2. The population distributions of NEC group and SIAT group

3.2   ECG Recordings

Each volunteer underwent 24-hour dynamic electrocardiogram monitoring.

3.3   Data Processing

The ECG Lab software provided by Hanix Company Ltd. was used in the processing and analysis of data in this experiment.

3.4   Analysis Parameters

A: Arrhythmia Indexes
Ventricular premature beat, Ventricular tachycardia, atrial tachycardia, Cardiac arrest were analyzed in this experiment.

B: Frequency Domain Indexes of HRV [14]
The frequency-domain indexes of the HRV signal, derived from its power spectral analysis, reflect autonomic cardiac modulation and were sensitive indicators of autonomic function in diabetic patients [16]. Five frequency domain measures were analyzed for each volunteer of each recording:
    TP: Total power
    ULF: Ultra Low Frequency

LF: Low Frequency .With the increase of sympathetic nerve activity, this index will increase.

HF: High Frequency .Which can be used as the quantitative index to detection cardiac vagal modulation activity level.

LF/HF: The ratio of LF and HF. Which is an index that relates the balance between sympathetic and parasympathetic parts of the autonomic nervous system and hence shows the level of sympathetic dominance.

C: Time Domain Indexes of HRV [6]

The HRV signal analysis provides a quantitative marker of the autonomic nervous system as the regulation mechanisms of HRV originate from the sympathetic and parasympathetic arms of the autonomic nervous system. Five time-domain measures were analyzed for each volunteer of each recording:

SDNN: main reflects the size of sympathetic and vagus nerve tension, used to assess the damage and recovery degree of overall of autonomic nervous system.

RMSSD, PNN50: mainly reflects the quick change back of heart rate variability, namely the size of tension of the vagus nerve.

SDANN: mainly reflects the slow change of heart rate variability, namely the size of tension of the sympathetic nerve.

SDNN index: main representative time path tension size of sympathetic or vagus nerve.

4   Results

4.1   The Results of Comparison and Analysis of Arrhythmia Indexes as Shown in Figure 3

Fig 3 manifested that the subjects in the SIAT group and NEC group are quite different. Though in some time period, the indexes of SIAT subjects might be higher than the NEC ones, from the general statistics view and considering the possible accidental error, the SIAT group subjects' indexes were lower than that of NEC group subjects. That means the average occurrence probability of various types of arrhythmia was higher in NEC group than the SIAT control group.

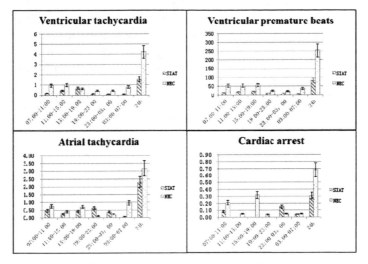

Fig. 3. Arrhythmia indexes comparisons

4.2   The Results of Comparison and Analysis of Time Domain Indexes of HRV as Shown in Figure 4

Fig 4 shows the frequency-domain indexes from the two groups. The SIAT group was 41% in TP indexes higher than the NEC group; for ULF indexes, the SIAT group was 27% higher than NEC group; the LF indexes were 11% and the HF indexes were 4%; in the ratio of LF and HF, the SIAT group was 0.5 larger than the NEC group. The results show that the control group—the SIAT group had a better performance in these frequency-domain indexes.

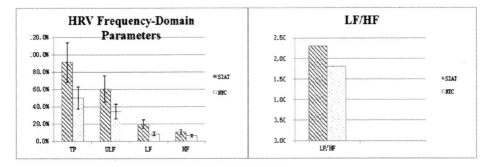

Fig. 4. Frequency-domain indexes comparisons

4.3   The Results of Comparison and Analysis of Time Domain Indexes of HRV as Shown in Figure 5

From Fig. 5, we found that average level of RMSSD and PNN50 indexes in two groups of people were basically equal. For SDNN-24 parameters, the range of SIAT group and NEC group were both within the range of 141±39ms; and for SDANN indexes, the ranges were within 127±35ms, there was none significant abnormality in the three parameters. As the ages of SIAT group (20~30) were lower than the NEC group (30~65), there was a higher SDNN indexes, SDANN indexes and SDNN parameters [17]. And these indexes reflected the sympathetic activity level, the bigger it was, the lower HRV rate was. So the occurrence probability of HRV in NEC group was higher than it in SIAT group.

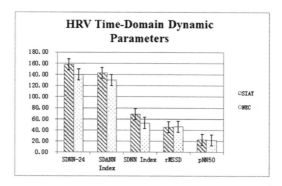

Fig. 5. Time-Domain Dynamic Parameters comparison

5   Conclusion

In this paper, we compared the dynamic ECG records from two groups of subjects - one in the very Northern of China (NEC group, from Shuangya Hill, Heilongjiang), one control group of SIAT (Shenzhen, Guangdong). Through the analysis and contrast, we can clearly find that average index of HRV and other features are markedly different in the experimental group which in the northeast of China with a special climate compared to the control group. The experiment results might be influenced by the living geographical environment, climate, diet habit, rest time and other lifestyle. From the arrhythmia indexes, there is a quite different in the subjects' performance, the results is accordant to the group selection, which subjects in NEC group were chosen in targets with angiocardiopathy symptoms. The subsequent analysis of frequency-domain indexes and time-domain indexes are quite different in explanation for the HRV analysis: frequency-domain indexes of the experimental group is quite different from the control group ones; the time-domain indexes were almost the same or have limited difference within the normal range. We infer that time-domain indexes might not be applicable for HRV analysis while the frequency-domain indexes are applicable.

Acknowledgement. This study was financed partially by the Projects of National Natural Science Foundation of China (Grant Nos. 60932001, 51105359 and 61072031), the National 863 Program of China (Grant No. 2012AA02A604), the

National 973 Program of China (Grant No. 2010CB732606) and the 'Low-cost Healthcare' Programs of Chinese Academy of Sciences, and the Guangdong Innovation Research Team Fund for Low-cost Healthcare. The experiment for acquiring the data is conducted in Hongxinglong, Shuangya City in Heilongjiang Province and Shenzhen with the assist of Hongxinglong Hospital and Shenzhen Institutes of Advanced Technology. Thanks Fu Wai Hospital for the assist of experiment.

References

1. Kelvyn, J., Moon, G.: Health, disease and society: an introduction to medical geography continued. Routledge & Kegan Paul Ltd. (1987)
2. Kass-Hout, T.A., Gray, S.K., Massoudi, B.L., et al.: NHIN, RHIOs, and public health. Journal of Public Health Management and Practice 13(1), 31–34 (2007)
3. Kleiger, R.E., Milier, J.P.: Deceased heart rate variability and association with increased mortality after acute myocardial infarction. Am. J. Cardiol. 59, 256 (1987)
4. Cripps, T.R., Malik, M., Camm, A.J.: Prognostic value of reduced heart rate variability after myocardial infarction: clinical evaluation of a new analysis method. Br. Heart J. 65, 14–19 (1991)
5. Wang, L., Fan, J., Zhang, Y., et al.: Low-Cost Health Care: Improving Care to Rural Chinese Communities Through the Innovations of Integrated Diagnostic Terminals and Cloud Computing Platforms. In: Technology Enabled Knowledge Translation for eHealth, pp. 401–412 (2012)
6. Rajendra Acharya, U., Paul Joseph, K., Kannathal, N., Lim, C.M., Jasjit Suri, S.: Heart Rate Variability. In: Advances in Cardiac Signal Processing, pp. 121–165. Springer, Heidelberg (2007)
7. Andrews, G., Cutchin, M., McCracken, K., et al.: Geographical Gerontology: the constitution of a discipline. Social Science & Medicine 65, 151–168 (2007)
8. Medical geography
9. Kearns, R.A.: Place and Health: Towards a Reformed Medical Geography. The Professional Geographer 139-147
10. Stein, P.K., Rich, M.W., Rottman, J.N., et al.: Stability of index of heart rate variability in patient with congestive heart failure. Am. Heart J. 129, 975–981 (1995)
11. Task force of the European society of cardiology and the North American society of pacing and electrophysiology. Heart Rate Variability standards of measurement, physiological interpretation, and clinical use. Circulation 93(5), 1043 (1996)
12. Pincus, A.M.: Approximate entropy as a measure of system complexity. Proc. Natt. Acad. Sci. USA 88, 2297–2301 (1991)
13. Smyth, F.: Medical geography: therapeutic places, spaces and networks. Progress in Human Geography 4, 488–495 (2005)
14. Niskanen, J.P., Tarvainen, M.P., Ranta-aho, P.O., Karjalainen, P.A.: Computer methods and programs in biomedicine. Software for Advanced HRV Analysis 1, 73–81 (2004)
15. Tarvainen, M.P., Ranta-aho, P.O., Karjalainen, P.A.: Biomedical Engineering. IEEE Transactions on, An advanced Detruding Method with Application to HRV Analysis 49, 172–175 (2002)
16. Lombardi, F.: Cardiac electrophysiology review. Clinical Implications of Present Physiological Understanding of HRV Components 3, 245–249 (2002)

Application of Analytic Hierarchy Process for User Needs Elicitation: A Preliminary Study on a Device for Auto-injection of Epinephrine

Leandro Pecchia[1], Jennifer L. Martin[1], Arthur G. Money[2], and Julie Barnet[2]

[1] Electrical Systems and Optics Research Division, University of Nottingham, UK
{leandro.pecchia,Jennifer.martin}@nottingahm.ac.uk
[2] Department of Information Systems and Computing, Brunel University
{arthur.money,Julie.Barnett}@brunel.ac.uk

Abstract. Understanding user needs is essential to design biomedical devices that are efficacious in real life (clinically effective). Few studies propose analytic quantitative methods to elicit user needs. This paper presents a preliminary application of the Analytic Hierarchy Process (AHP) to elicit user needs. As a case study we focused on the use of a biomedical device for auto-injection of epinephrine to treat severe allergic reactions. Although the study presented is on-going, the methods we describe provide valuable insights into how quantitative methods can be applied to user needs elicitation.

Keywords: Analytic Hierarchy Process, anaphylaxis, epinephrine auto-injector, user needs, health technology assessment, medical device design.

1   Introduction

The juxtaposition of economic and clinical evaluation raises new issues in the design of clinical trials. Currently pivotal phase III trials are designed to test safety and efficacy (does the drug work under optimal circumstances?) and not to answer questions about the effectiveness of a drug, which is the more relevant question for health technology assessment (does the drug work in usual care?) [1]. Failure to effectively collect and consider user needs during development has been shown to reduce the real world efficacy (clinical effectiveness) of medical devices, In the literature there are several studies that attempt to address this issue. The majority of these studies use qualitative research methods to investigate why promising health technologies fail in the real world. These studies can provide valuable insights to inform the design of new healthcare services or products; however, often their diffusion among manufactures is limited. One of the reasons is that the results of the majority of these studies are presented as qualitative reports, in a form that designers, and engineers find difficult to appraise and incorporate into the design process [2].

This paper presents an application of a quantitative method to elicit user needs of epinephrine auto-injectors (EAIs) for the treatment of anaphylaxis. In clinical trials these devices have been shown to perform well, reducing mortality, morbidity and

G. Huang et al. (Eds.): HIS 2013, LNCS 7798, pp. 258–264, 2013.

hospitalization [3]. However, research has also shown that the effectiveness of EAIs is often reduced because patients do not always carry the device with them at all times for reasons other than technical ones [4,5].

The use of *scientific quantitative methods to support decision making* is considered necessary in healthcare organizations, where the personnel are committed to follow only the best available evidence according to well-designed trials [6], meta-analyses [7] or network meta-analyses [8]. Nonetheless, despite the hierarchy of evidence, the complexities of medical device decision-making require a spectrum of qualitative and quantitative information [9]. The method proposed in this study to quantify user needs is an adaptation of Analytic Hierarchy Process (AHP) [10], which is a multi-dimensional, multi-level and multifactorial decision-making method already used in medicine [11] and in Health Technology Assessment (HTA) studies [12]. AHP aims to prioritize decisional variables, based on three main ideas: (1) grouping all the variables into meaningful categories and sub-categories, (2) structuring the problem as a decisional tree; (3) performing pairwise comparisons between needs (leafs of the tree from the same node) and needs' categories (tree nodes); defining a coherent framework of quantitative and qualitative knowledge This hierarchical approach allows the construction of a consistent framework for step-by-step decision-making, breaking a complex problem down into a number of smaller, less-complex ones, which decision makers can more easily deal with. This is particularly useful when the decision-makers (in this study, patients) are not familiar with complex decision making methods [13]. At the moment of preparing this paper, both tree and questionnaires are being piloted among researchers experienced in user needs elicitation and with a selected group of EAI users who have already participated in previous user needs assessment. A final version of a 'tree-of-needs' and a corresponding questionnaire is expected to be submitted to a wider number of users in the next three months.

In this paper, after describing the case study, the method is introduced and the preliminary version of the tree-of-needs and questionnaires are presented.

2   Case Study

Anaphylaxis is a life threatening allergic reaction which affects the respiratory and/or cardiovascular systems [14]. A key component in the treatment of anaphylaxis relies on the patient providing routine self-care and management to prevent this occurring [15]. Whilst anaphylaxis may be triggered by exposure to latex rubber, insect venom and medication, the most common cause is exposure to foods including peanuts, nuts, fish, milk and eggs [16]. The incidence of anaphylaxis has risen dramatically in recent years, as reflected by a sevenfold increase in anaphylaxis-related UK hospital admission between 1990/1 and 2003/4 [17]. The treatment of anaphylaxis is a prompt intramuscular injection of epinephrine, typically administered by the patient themselves. It is therefore not surprising that prescriptions of EAIs have risen, with 10,700 prescriptions being issued in England in 2001, rising to 21,100 in 2005 [18]. Patients considered at risk of anaphylaxis are prescribed at least one EAI, which in accordance with self-care best practice for this condition, is to be carried by the patient at all times so that the device is readily available for rapid self-treatment when necessary [19]. It is widely accepted that not having an EAI available at the scene of a severe anaphylaxis event puts the patient at significant risk of a fatal outcome [5].

Although EAIs have been designed to be used as self-treatment devices by patients there is evidence to suggest that patients often do not engage in appropriate self-care practices such as the carriage and use of the device when necessary [4]. A study of fatal anaphylactic reactions revealed that only 10% of individuals actually had epinephrine to hand when it was required [20], and even when the device is to hand, it is often not used [21].

Despite the serious consequences of not having such a device to hand, there is a lack of research that considers the experiences and attitudes of patients and the strategies they use in the delivery of care for this condition. More specifically, to the best of our knowledge, little research has been carried out to specifically explore, from the adult patients' perspective, what patient motivations are for carriage or non-carriage of EAIs and/or their deployment/non-deployment at appropriate times [22].

3   Method

The following nine steps describe the whole method of this study, although at this moment the steps 4-9 are not yet completed:

1. User needs identification. This step directly involved 2 domain experts (JB and AM) and fifteen device users (interviewed in a previous study).
2. Design of a tree-of-needs with nodes (categories), sub-nodes (sub-categories) and leaf (needs). This involved 1 expert of user needs elicitation (JM) and one researcher experienced in the AHP (LP).
3. Development of questionnaires (1 AHP expert).
4. Tree piloting. This involves "n" domain expert/s, where n is dynamically established according to variability among experts' opinions.
5. Questionnaire piloting. This involves a small group of selected responders (from 3 to 5), with experience of participation in user needs elicitation studies.
6. Final tree and questionnaire development (1 AHP expert, 1 domain expert and 1 experienced elicitor).
7. Final questionnaire submission.
8. Data analysis end results presentation (1 expert of AHP)
9. Results interpretation and discussion (1 expert of AHP, domain expert, 1 elicitor and some users).

The AHP method is applied following the workflow represented in the Figure 1. Details about AHP methods can be found in other papers [23-25], which are freely accessible (http://eprints.nottingham.ac.uk/). We do not describe the method in detail as no further modification was proposed in this study. In the following sections we describe the methodological details that are relevant to this study.

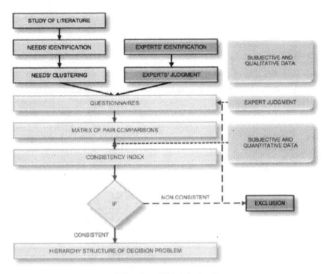

Fig. 1. AHP Method

3.1 Questionnaires Design

The questionnaire was designed to minimize possible responder bias. For instance, each element was presented the same number of times on the top left and the right, at the top and at the bottom of the questionnaire as responders writing from left to right and top-down can be more likely to judge the elements on top-left as more important than those bottom right. Moreover, the sequence of comparisons (A with B, B with C and C with A) was adapted to minimize intransitive judgments and no more than 4 elements were included into each category, to ensure the number of questions included in each questionnaire was not overwhelming for the user. Finally just 5 levels of the Saaty scale were used in order to facilitate responder consistency. The questionnaire was first piloted in the lab and then with a small group of patients, who had already participated in previous studies.

ACCORDING TO YOUR PERSONAL EXPERIENCE, HOW IMPORTANT IS EACH FACTOR
ON THE LEFT COMPARED WITH EACH ONE ON THE RIGHT IN CHOOSING NOT TO CARRY THE EPIPEN?
REGARDING THE "SOCIAL ISSUES", AND ESPECIALLY "OTHER PEOPLE OPINION", IN YOUR EXPERIENCE:

HOW PEOPLE PERCEIVE YOU	IS	MUCH MORE	MORE	EQUALLY	LESS	MUCH LESS	IMPORTANT THAN	HOW PEOPLE PERCEIVE THE EPIPEN
HOW PEOPLE PERCEIVE THE EPIPEN	IS	MUCH MORE	MORE	EQUALLY	LESS	MUCH LESS	IMPORTANT THAN	HOW PEOPLE PERCEIVE YOUR CONDITION
HOW PEOPLE PERCEIVE YOUR CONDITION	IS	MUCH MORE	MORE	EQUALLY	LESS	MUCH LESS	IMPORTANT THAN	HOW PEOPLE PERCEIVE YOU

Fig. 2. Example of questionnaire

4   Results

The preliminary tree-of needs that was developed as a result of carrying out this process is presented in Figure 2.This tree was used to develop the questionnaire according to the layout presented in Figure 3.

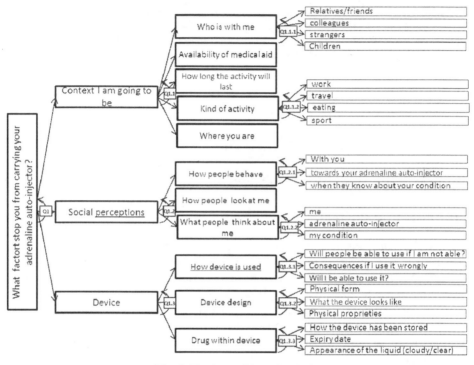

Fig. 3. Needs tree hierarchy design

The hierarchy and the questionnaire are currently undergoing piloting among other researchers and a selected group of patients who have already participated in previous user needs assessment. The final version of the questionnaire is expected to be completed and analyzed in the next three months. Nonetheless, the process of creating the hierarchy illustrates how qualitative data can be used as a basis for a quantitative investigation into user needs. The next steps of this study will also provide qualitative information about the relative importance of each need compared to all the others. This will allows experts, and especially manufacturers, to focus on those needs that users considered more important.

5   Conclusion

In this paper, an application of the AHP to elicit user needs was presented. This is an important issue in HTA as the user needs that affect the efficacy of biomedical devices in the real word (effectiveness) are often underestimated. Although a growing sensitivity to these issues has recently been demonstrated, few studies have proposed analytic methods to investigate them. The preliminary results presented here (mainly the tree-of-needs) have already provided a useful and interesting way of analyzing, synthesizing and transforming user data which were collected during previous qualitative studies. This paper demonstrates how AHP can be used to transform user data into a form that can be more easily incorporated into medical device design. It is hoped that this method will provide a means of bridging the gap that has been shown to exist between the broad and rich nature of user research and the specific and focused requirements of manufacturers when using this data to produce design specifications. Finally, qualitative information about the priorities of needs will complement the quantitative insight.

Acknowledgements. LP, JM, JB acknowledge support of this work through the MATCH Programme (EPSRC Grant EP/F063822/1) although the views expressed are entirely their own.

References

1. Bombardier, C., Maetzel, A.: Pharmacoeconomic evaluation of new treatments: efficacy versus effectiveness studies? Annals of the Rheumatic Diseases 58(suppl. 1), 182–185 (1999)
2. Martin, J.L., Barnett, J.: Integrating the results of user research into medical device development: insights from a case study. BMC Medical Informatics and Decision Making 12, 74 (2012), doi:10.1186/1472-6947-12-74
3. Simons, K.J., Simons, F.E.R.: Epinephrine and its use in anaphylaxis: current issues. Current Opinion in Allergy and Clinical Immunology 10(4), 354–361 (2010), doi:10.1097/Aci.0b013e32833bc670
4. Cummings, A.J., Knibb, R.C., Erlewyn-Lajeunesse, M., King, R.M., Roberts, G., Lucas, J.S.A.: Management of nut allergy influences quality of life and anxiety in children and their mothers. Pediat Allerg Imm-Uk 21(4), 586–594 (2010), doi:10.1111/j.1399-3038.2009.00975.x
5. Pumphrey, R.: When should self-injectible epinephrine be prescribed for food allergy and when should it be used? Current Opinion in Allergy and Clinical Immunology 8(3), 254–260 (2008), doi:10.1097/ACI.0b013e3282ffb168
6. Bracale, U., Rovani, M., Picardo, A., Merola, G., Pignata, G., Sodo, M., Di Salvo, E., Ratto, E.L., Noceti, A., Melillo, P., Pecchia, L.: Beneficial effects of fibrin glue (Quixil) versus Lichtenstein conventional technique in inguinal hernia repair: a randomized clinical trial. Hernia (2012), doi:10.1007/s10029-012-1020-4
7. Bracale, U., Rovani, M., Bracale, M., Pignata, G., Corcione, F., Pecchia, L.: Totally laparoscopic gastrectomy for gastric cancer: Meta-analysis of short-term outcomes. Minim. Invasive Ther. Allied Technol. 21(3), 150–160 (2011), doi:10.3109/13645706.2011.588712
8. Bracale, U., Rovani, M., Melillo, P., Merola, G., Pecchia, L.: Which is the best laparoscopic approach for inguinal hernia repair: TEP or TAPP? A network meta-analysis. Surgical Endoscopy 26(12), 3355–3366 (2012), doi:10.1007/s00464-012-2382-5

9. Leys, M.: Health care policy: qualitative evidence and health technology assessment. Health Policy 65(3), 217–226 (2003), doi:10.1016/S0168-8510(02)00209-9

10. Saaty, T.L., Vargas, L.G.: Models, methods, concepts & applications of the analytic hierarchy process. International series in operations research & management science, vol. 34. Kluwer Academic Publishers, Boston (2001)

11. Liberatore, M.J., Nydick, R.L.: The analytic hierarchy process in medical and health care decision making: A literature review. European Journal of Operational Research 189(1), 194–207 (2008), doi:10.1016/j.ejor.2007.05.001

12. Pecchia, L., Craven, M.P.: Early stage Health Technology Assessment (HTA) of biomedical devices. In: Long, M. (ed.) World Congress on Medical Physics and Biomedical Engineering. IFMBE Proceedings, vol. 39, pp. 1525–1528. Springer, Heidelberg (2013)

13. Dolan, J.G.: Are Patients Capable of Using the Analytic Hierarchy Process and Willing to Use It to Help Make Clinical Decisions. Medical Decision Making 15(1), 76–80 (1995), doi:10.1177/0272989x9501500111

14. Muraro, A., Roberts, G., Clark, A., Eigenmann, P.A., Halken, S., Lack, G., Moneret-Vautrin, A., Niggemann, B., Rance, F., Tfac, E.: The management of anaphylaxis in childhood: position paper of the European academy of allergology and clinical immunology. Allergy 62(8), 857–871 (2007), doi:10.1111/j.1398-9995.2007.01421.x

15. Choo, K., Sheikh, A.: Action plans for the long-term management of anaphylaxis: systematic review of effectiveness. Clin. Exp. Allergy 37(7), 1090–1094 (2007), doi:10.1111/j.1365-2222.2007.02711.x

16. Ewan, P.W.: ABC of allergies - Anaphylaxis. Brit. Med. J. 316(7142), 1442–1445 (1998)

17. Gupta, R., Sheikh, A., Strachan, D.P., Anderson, H.R.: Time trends in allergic disorders in the UK. Thorax 62(1), 91–96 (2007), doi:10.1136/thx.2004.038844

18. Sheikh, A., Hippisley-Cox, J., Newton, J., Fenty, J.: Trends in national incidence, lifetime prevalence and adrenaline prescribing for anaphylaxis in England. Journal of the Royal Society of Medicine 101(3), 139–143 (2008), doi:10.1258/jrsm.2008.070306

19. Baral, V.R., Hourihane, J.O.: Food allergy in children. Postgraduate Medical Journal 81(961), 693–701 (2005), doi:10.1136/pgmj.2004.030288

20. Bock, S.A., Munoz-Furlong, A., Sampson, H.A.: Fatalities due to anaphylactic reactions to foods. J. Allergy Clin. Immun. 107(1), 191–193 (2001), doi:10.1067/mai.2001.112031

21. Simons, F.E.R.: First-aid treatment of anaphylaxis to food: Focus on epinephrine. J. Allergy Clin. Immun. 113(5), 837–844 (2004), doi:10.1016/j.jaci.2004.01.079

22. Money, A.G., Barnett, J., Kuljis, J., Lucas, J.: Patient perceptions of epinephrine autoinjectors: exploring barriers to use. Scandinavian Journal of Caring Sciences (2012), doi:10.1111/j.1471-6712.2012.01045.x

23. Pecchia, L., Bath, P.A., Pendleton, N., Bracale, M.: Web-based system for assessing risk factors for falls in community-dwelling elderly people using the analytic hierarchy process. International Journal of the Analytic Hierarchy Process 2(2), 135–157 (2010)

24. Pecchia, L., Bath, P.A., Pendleton, N., Bracale, M.: Analytic Hierarchy Process (AHP) for Examining Healthcare Professionals' Assessments of Risk Factors The Relative Importance of Risk Factors for Falls in Community-dwelling Older People. Methods Inf. Med. 50(5), 435–444 (2011), doi:10.3414/Me10-01-0028

25. Pecchia, L., Martin, J.L., Ragozzino, A., Vanzanella, C., Scognamiglio, A., Mirarchi, L., Morgan, S.P.: User needs elicitation via analytic hierarchy process (AHP). A case study on a Computed Tomography (CT) scanner. BMC Med. Inform. Decis. Mak. 13(1), 2 (2013), doi:10.1186/1472-6947-13-2

Mathematical-Morphology-Based Edge Detection of Retinal Vessels in Retinal Images

Zhangwei Jiang[1], Yanchun Zhang[2], Jing He[2], Guangyan Huang[2],
and Jing Yang[1]

[1] GUCAS-VU Joint Lab for Social Computing and E-Health Research,
University of Chinese Academy of Sciences, China
[2] Center for Applied Informatics, School of Engineering and Science,
Victoria University, Australia

1 Introduction

Of all the important small and medium-sized blood vessels of the human body, retinal blood vessels are the only deep capillary that can be directly observed by a non-traumatic method. Retinal vascular morphology, such as vessel diameter, shape and distribution, is influenced by systemic diseases (Martinez-Perez, Hughes, Thom and Parker 2007). We can use digital fundus photography and analysis of retinal vascular morphology to find the relationship between the changes in vascular morphology and diabetes for the diagnosis of diseases. We aim at developing a retinal image processing system, that can analyze retinal images and provide helpful information for diagnosis.

In this demonstration we present a retinal image processing system which we implement the morphology method and other image processing technologies. The preprocessing will be conducted before the retinal segmentation, and then the retinal features will be detected based on retinal vessels, which can be mocked up as a model and used as classification criteria to determine whether the input retinal image is normal. A novel algorithm using an improved morphological edge detector and self-adaptive weighted synthesis based on information entropy is introduced into the system.

In the rest of this demo paper, we first describe, in short, several concepts used in our system. Then, we demonstrate the system architecture.

2 Image Enhancement and Segmentation

Common image pre-processing skills are image enhancement, including spatial domain enhancement and frequency domain enhancement (Rafael and Richard, 1997). Spatial domain enhancement contains point operations and neighborhood operations. Point operations map the input image to the output image, where the gray value of each pixel of the output image is only determined by the corresponding pixel of an input image, containing gray scale correction, grey level transformation and histogram correction. The purpose of point operations is to make the imaging uniform, or to expand the dynamic range and contrast.

G. Huang et al. (Eds.): HIS 2013, LNCS 7798, pp. 265–268, 2013.

Neighborhood operations map the input image to the output image, where the gray value of each pixel of the output image is determined by the corresponding pixel and its neighborhood pixels of the input image. The neighborhood operations can be divided into smoothing and sharpening. Smoothing methods contain mean filter and median filter. Mean filter uses the mean gray value to replace the pixel values of the original image, and median filtering uses the median gray value. The purpose of image sharpening is to enhance the image contour. Edges refer to the pixels whose gray values have step change or roof shape change. Edge detection methods include the Roberts operator, Prewitt operator, Sobel operator and direction operator.

Image segmentation is a process of segmenting the image into several disjoint connected regions. Existing image segmentation algorithms include thresholding methods, region growing methods, clustering methods (Mohammad and Ali 2002). The thresholding method is a relatively simple method where the threshold is used to distinguish the object's pixels from the background. Pixels larger than or equal to the threshold value belong to the object, and the others belong to the background. The threshold can be divided into global threshold, adaptive threshold and optimum threshold. A global threshold is a uniform threshold value for all pixels in segmentation processing and it works well for the images where there is significant difference between the object and the background. In some cases, objects in different regions have different contrast with the background which will need adaptive thresholding. The threshold value of the optimum threshold method is obtained by the analysis of the histogram of the image which is appropriate for specific tasks. The region growing methods select a point as the seed and calculate the level of similarity of gray or texture between the seed pixel and its surrounding pixels, then merge the adjacent pixel with similar features into the current region. Clustering methods classify the pixels into different classes with a certain clustering algorithm based on the features of gray or texture.

3   Advanced Edge Detection Algorithm

In general morphological edge detection algorithms using simple symmetric structure elements (such as disc, square, cross), which are sensitive to the edges that have the same direction with the structure elements, can weaken the sensitivity of the edges that have different directions with the structure elements (Zhao, Gui and Chen, 2006). But the multi-structure element covers almost all directions, therefore, it can detect the edges of different directions. The traditional morphology method for the synthesis is mean synthesis. Although it can obtain a good edge, it does not give full play to the anti-noise performance of the large-scale structure element. Therefore, this paper presents a new morphological edge detection algorithm, a method using the characteristic of the edge embodied by the entropy, which can adaptively determine the weighted value for weighted synthesis to eventually obtain a better edge.

4   Results

Our source image is a retina image whose format is BMP and our development environment is VC++ 6.0.

Our fundus image processing and analysis system includes a fundus image pre-processing module, image enhancement module, vascular extraction module, and image diagnosing module, represented as follows:

(1) Fundus image pre-processing module: the main function is to realize basic operations of image processing, including template operations, geometric transformations, denoising and brightness contrast improvement, as shown in Figure 1.

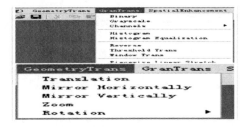

Fig. 1. Fundus image pre-processing module

(2) Fundus image enhancement module: image enhancement falls into two parts, spatial domain enhancement and frequency domain enhancement, which includes grayscale correction, image sharpening, smoothing and so on. The grayscale correction includes gray transformation and histogram correction. Image smoothing is mainly image filtering, including mean filtering, median filtering and Gaussian matched filtering, as shown in Figure 2.

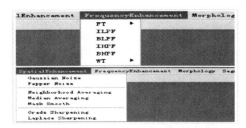

Fig. 2. Fundus image enhancement module

(3) Retinal vascular extraction module: This module implements the function of extracting blood vessels in the fundus image. In this part, we realize the traditional morphology method and our advanced method with multi-structure elements and weighted synthesis, as shown in Figure 3.

(4) Diagnosing module: In this module we calculate several features of the retinal image and then accordingly judge whether it is normal, as shown in Figure 4.

Fig. 3. Retinal vascular extraction module

Fig. 4. Diagnosing module

5   Conclusion and Future Work

In this demonstration, we presented the retinal image processing system that implemented our advanced algorithm based on the morphology method and several other image processing technologies. Compared with traditional methods, our algorithm has better performance in terms of definition and anti-noise. Furthermore, our system is able to distinguish pathological retinal images. We are currently working on improving the software interface and diagnostic accuracy so that it can be applied to practice.

References

1. Martinez-Perez, M.E., Hughes, A.D., Thom, S.A., Parker, K.H.: Improvement of a retinal blood vessel segmentation method using the Insight Segmentation and Registration Toolkit (ITK). In: Proc. IEEE 29th Annual Int. Conf. EMBS, Lyon, France, pp. 892–895 (2007)
2. Rafael, C.G., Richard, E.W.: Digital Image Processing, 2nd edn. PHEI, Beijing (2011) (Ruan, Q., Ruan, Y., trans.) (Original work published 1977)
3. Mohammad, S.M., Ali, M.: Retinal Image Analysis Using Curvelet Transform and Multi-Structure Elements Morphology by Reconstruction. IEEE Tran. (2002)
4. Zhao, Y., Gui, W., Chen, Z.: A Multi-structure Elements Algorithm for Image Edge-detection Based on Mathematical Morphology. Intelligent Control and Automation (2005)

Online Action Recognition
by Template Matching

Xin Zhao[1,2,3], Sen Wang[2,3], Xue Li[2], and Hao Lan Zhang[1]

[1] NIT, Zhejiang University, 1 Xuefu Road, Ningbo, China
[2] School of ITEE, The University of Queensland, Australia
[3] The Australian E-Health Research Centre, CSIRO, Australia

Abstract. Human action recognition from video has attracted great attentions from various communities due to its wide applications. Regarded as an effective way to analyze human movements, human skeleton is extracted and represents human body as dots and lines, Recently, depth-cameras make skeleton tracking become practical. Based on the extraction and template matching, we develop a system for online human action segmentation and recognition in this paper. We proposed a method to generate action templates that can be used to represent intra-class variations. We then adopted efficient subsequence matching algorithm for online process. The experimental results demonstrated the effectiveness and efficiency of our system.

Keywords: Action recognition, Subsequence matching, Depth-camera.

1 Introduction

In this paper, we create a system which can recognize indoor human actions from video and record these actions as form of text in real time. This technique can be easily applied in smart homes and nursing cares. For example, the elders' abnormal actions can be detected in real time and can be monitored/displayed from their children's cell phones.

In our system, we use human skeletons tracked from video to recognize actions represented with these skeletons. Skeleton of human body is represented as dots and lines. Dots represent human joints, and lines represent human body parts. As Johansson [1] suggested, skeleton itself is sufficient to distinguish different types of human actions. The major advantage of skeleton is that it can be used to analyze detailed levels of human movements [2]. Furthermore, the recent introduction of cost-effective depth camera and the related motion capturing technique [3] enable ot track human skeleton efficiently. This phenomenon has brought on a new trend of research on human skeleton based action recognition [4].

In human action recognition from videos, four kinds of intra-class variations may affect the effectiveness: viewpoint, anthropometry, execution rate, and personal style. Viewpoint variation describes the relationship between actor and viewpoint of camera. Anthropometry variation is related to the size of actor,

G. Huang et al. (Eds.): HIS 2013, LNCS 7798, pp. 269–272, 2013.

which refers human physical attributes and doesn't change with human movements. Execution rate variation is related with the temporal variability caused by actor or by differing camera frame-rates. Personal style is also required to be considered as people may perform same action in different styles.

Furthermore, online recognition from unsegmented stream is another challenge. Because streaming data does not provide pre-segmented instances, we need to find a way to segment a right number of frames for recognition. There are two kinds of traditional methods for segmentation. The first method uses fixed-size sliding window technique [4] [5]. Each segment is treated as one instance. Then machine learning techniques are used to do classification. Unfortunately, fixed-size sliding window suffers poor performance when there is execution rate variation. The second method matches pre-defined action templates from stream. Each template represents one type of action. Each matched subsequence treated as one instance and assigned with the same action label as the corresponding template. Sakurai et al in [6] proposed an algorithm to solve the problem of efficient subsequence matching in stream and showed the potential of their algorithm to handle the problem of human action recognition. However, they simply manually chose one segmented instance as the template of this type of action. This template cannot represent personal style variation.

In this paper, we develop a system for online human action recognition. We proposed a method to generate action templates which can represent these four kinds of intra-class variations. The algorithm proposed in [6] is adopted for efficient subsequence matching.

2   Methodology

Our system can be divided into three parts: skeleton preprocessing to generate features for recognition, template learning to obtain templates for matching, subsequence matching to give recognition results: action labels of each frame in stream.

Skeleton preprocessing. In this system, we use the skeleton model of openni developed by Primesense and the depth-camera kinect released by Microsoft. The skeleton of human body is a tree structure. Each node in the tree represents one joint position associate with 3 coordinates (x, y, z), and each edge between two connected nodes means one body part. There are 14 joints and 13 body parts in openni skeleton model. Root joint is specified as the joint "spine" s . Body axis a is defined as the line from "spine" to the middle point of "left shoulder" l and "right shoulder" r, i.e., $a = \frac{(l+r)/2-s}{||(l+r)/2-s||}$. The body orientation o is the normal vector of triangle with vertices "spine", "right shoulder" and "left shoulder", i.e., $o = \frac{(s-l)\times(l-r)}{||(s-l)\times(l-r)||}$, where symbol '$\times$' is cross product of vectors.

For human action recognition, joint positions should be transformed into joint angles, which are viewpoint invariant and anthropometry invariant. Firstly, we transform joint positions into the normalized coordinate system, where $s = (0,0,0)$ and $a = (0,1,0)$ and $o = (0,0,-1)$ after transformation. Then we transform joint positions into joint angles. Assume p and q are two connected joints.

p is the parent of q. Joint angle of q is $\frac{q-p}{||q-p||}$. Angle of root joint is omitted. For simplification, we use skeleton data to refer to both joint positions and joint angles in this paper.

Template learning. We learn one template with pre-segmented instances of one action type. One template is consisted with two sequences: one is average sequence which represents standard movement, and the other is deviation sequence which represents personal style variation. We compute the optimal alignment of each instance with initial template by applying Dynamic Time Warping (DTW) distance to eliminate execution rate variation. Initial template is one instance randomly selected. According to these optimal alignments, the instances are locally stretched or contracted, where time stretching is simulated by duplicating columns, while time contractions are resolved by forming a average of the columns. These instances after alignment and the initial template are used to compute template. The average over these instances is the average sequence of template, the deviation over these instances is the deviation sequence of template.

The left curves in Figure 1 illustrates action templates. The green skeletons are the average ones. The red range on each joint illustrates the deviations. The templates of curves from top to bottom are actions "kick with right leg" ("kickSideR" for short), "wave" and "walk" respectively.

Subsequence matching. In order to match a template from unsegmented stream, we use a variant of DTW. Let $X = (x_1, x_2, ..., x_n, ...)$ be skeleton data stream, where x_n is the most recent skeleton data and n increases with every new time-tick. $T = (A; B)$ is a template of one action type. We aim to identify the optimal subsequence of X with ending position at current time-tick n. Around all subsequences of X with ending position at n, the DTW distance between template and optimal subsequence is the minimum. The stream should be treated online fashion. Inspired by [6], we calculate optimal subsequence according to each template at every time-tick in stream. At time-tick n, the frame is identified as the action type of corresponding template which obtains minimal distance and the distance is smaller than the given threshold ϵ.

Figure 1 shows an example of template matching from stream. Axis X represents time-ticks of skeleton data stream, and axis Y represents the DTW distances between optimal subsequences and templates. The recognized results are shown on top, which consist with the skeleton data stream. The red lines represent the threshold ϵ.

3   Outcomes

Datasets. We capture one skeleton stream performed by one subject. There are 1000 frames and three types of actions: "kickSideR", "wave", "walk". We choose 3 instances of each type of action to learn its template. We capture another skeleton stream performed by another subject for evaluation. This stream consists of 1000 frames and contains 6 "kickSideR", 6 "wave", 43 "walk" instances. The ground truth is manually labeled.

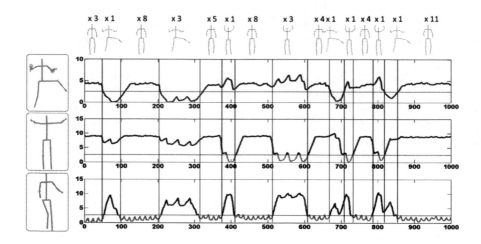

Fig. 1. Example of template matching from stream.

Effectiveness. The classification result of one instance is decided by the majority frames with same label. The threshold ϵ is set to 2.5. As shown in Figure 1, all instances are correctly classified.

Efficiency. We preform our experiments with hardware of "i7 860 CPU" and "4G RAM", and softwares of Matlab hybrid with parts of C code. With our approach, more than 30 frames can be processed per second.

Acknowledgments. The work reported in this paper is partially supported by Ningbo Natural Science Foundation (No. 2012A610025, No.2012A610060), Ningbo Soft Science Grant (No. 2012A10050), Ningbo International Cooperation Grant (No. 2012D10020) and the National Natural Science Fund of China (No. 71271191, No 61272480).

References

1. Johansson, G.: Visual motion perception. Scientific American (1975)
2. Aggarwal, J.K., Ryoo, M.S.: Human activity analysis: A review. CSUR 43, 16 (2011)
3. Shotton, J., Fitzgibbon, A.W., Cook, M., Sharp, T., Finocchio, M., Moore, R., Kipman, A., Blake, A.: Real-time human pose recognition in parts from single depth images. In: CVPR (2011)
4. Fothergill, S., Mentis, H.M., Kohli, P., Nowozin, S.: Instructing people for training gestural interactive systems. In: CHI (2012)
5. Zhao, X., Li, X., Pang, C., Wang, S.: Human action recognition based on semi-supervised discriminant analysis with global constraint. Neurocomputing (2012)
6. Sakurai, Y., Faloutsos, C., Yamamuro, M.: Stream monitoring under the time warping distance. In: ICDE (2007)

User-Centred ICT-Design for Occupational Health: A Case Description

Marjo Rissanen

Aalto University, School of Science, Finland
mkrissan@gmail.com

Abstract. In workplace health promotion there is a continuous need for adaptive strategies, tools, methods, and wellness programmes. Likewise, there are also needed abilities to mix different approaches in training. An application was produced to prevent neck and shoulder disorders and to enhance general well-being at work. Its purpose was to support especially blended health programmes. Nowadays, there are many applications targeted for self-paced use to also manage these kinds of symptoms. However, because a tutor-based consultative training still has its place in this area, innovative ICT-applications are also needed to support blended programmes or a tutor-driven instruction in occupational health. Such applications are needed in such sub-specialties and focus environments where a need for user-centred and customized health programmes is obvious.

Keywords: ICT-design, occupational health.

1 Introduction

Occupational health is a meaningful focus area for innovative ICT-production because maintaining a good mental and physical work condition is cost-effective. Innovative eHealth applications should manifest product quality which is useful and inspiring and also offers appropriate process quality for an occupational tutored or self-paced training. Customer quality means that the application takes into account the different user needs. Individual health problems and symptoms may differ even in such a case when the diagnosis in question might be common which makes user-centred design in this field challenging. This paper is a description of an application suitable for training which combines classical class interventions and self-paced learning. Design research type of an emphasis with its phase which concentrates on the principles of the artefact and its training environment takes place as an approach.

2 Quality in the Environment and Toolkit Integration

At the occupational field the training process may consist of different phases; individual tutoring, group sessions, and distant learning (e.g., assisted with delivered on-line material). Blended programmes utilize typically offline and online training phases.

G. Huang et al. (Eds.): HIS 2013, LNCS 7798, pp. 273–276, 2013.
© Springer-Verlag Berlin Heidelberg 2013

However, a totally self-paced learning in knowledge intensive areas may be demanding. As such, health related personal problems may be challenging and ill-structured. The offered information content may be too wide or may contain too detailed or too superficial information in relation to individual needs. One must be able to analyze, synthesize, and evaluate the given information when applying it which resembles self-diagnostics. Also applications with well functioning online educator systems do not always replace a traditional face-to-face contact. For this reason, the developed application was targeted to support blended programmes.

3   Description of the Application

3.1   Motivational Aspects

Neck and shoulder disorders are typical troubles in many work sectors. The etiology of neck pain contains physical, psychosocial, and individual factors [1]. Poor lifestyle habits, physical work overload, poor ergonomic facilities, and also stress related factors may cause musculoskeletal problems, neck tension, related symptoms, and illnesses. For the prevention and cure supervised exercise and manual therapy e.g., are proved useful as treatments [2] but further research is needed to develop appropriate primary prevention strategies [3]. The purpose of the application is to enhance wellbeing at work, be helpful, and prevent neck and shoulder disorders. User groups consist of persons who work in areas where these disorders cause many sickness days and diminish the work condition (e.g., office workers, IT-workers, musicians; generally unbalanced physical activities).

Fig. 1. Video for self-paced learning: tips for stretching (prevention with exercise)

3.2   Utilization Ideology and the Features of the Application

The product gives sinew for a classical classroom type instruction which is supported with employees' self-learning periods. Some informative areas are as such suitable for self-learning (e.g., knowledge of etiology, symptoms, curative possibilities etc.). Also, guiding videos giving tips for training, exercise, and stretching (Fig. 1) and self-evaluative tests represent such elements which the user can manage and learn by one-self. However, some theoretical, knowledge intensive parts and individual needs can be handled more specifically with the aid of a coach. Also, the co-operation with the interest units of the company ensures better commitment. A Macromedia Flash based application consists of the following components: text, audio, illustrations, video clips, supporting tools for interactive sessions (topics and task specific forms for self-tests and aids for discussions). The main modules are: neck and shoulder disorders and their prevalence, managing stress factors in work, health promotion practices at work, environmental factors and ergonomics, preventive exercising, and the value of rest and sleep (see Fig. 2). Each module has further subcategories. Because the application consists of the different emphasis areas around the theme, the connective training programme can form a training entity, in which each consultation day concentrates on a certain sub-theme area. Such an ideology ensures that the training programme is not too multifaceted and for that reason too demanding to capture and manage at the same time.

Fig. 2. Main page lists the areas of substance and extended health dimensions

3.3   Innovativeness and Usefulness to Practice

There are applications which are planned for self-paced learning and for such applications there is naturally also a continuous need in the occupational health sector. Also this application contains parts which are suitable for self-paced learning. However, a blended training model gives flexibility for areas where there exists much variation among the symptoms, disorders, and connected questions. The professional tutor can complement informative parts according to the situational needs ensuring this way more contextualized and customized programmes. Tutor's consultative support gives more security (e.g., under-exercising or over-exercising are both undesirable situations). The coach is a controller of connected discipline techniques and strategies with a purpose to increase appropriate behaviour models. A tutor-driven programme offers co-ordination ensuring a more homogenous information level and better adaptation to the organizational context.

4   Conclusions

When the mission of digital informative products is pronounced clearly enough it gives empowerment for the projects and connected product design. When the product is an essential part in framing the training entity, every training programme gives more cues for coaches; how to apply the toolkit in the training in the best possible way and how to maturate its featuring. The engagement with users, design dialogue, helps to identify user needs, opinions, and expectations which are critical to deployment of the products [4, 5] and can ensure that *"an innovation is not evaluated and understood in isolation, but rather as an integral part of the context"* [6]. A user-centred design gives much attention to the opinions and value aspects of intended user groups but their integration for the design is not always the answer for all questions: e.g., in occupational health user groups are heterogeneous but also different work environments may have different needs and demands. Therefore, well fitted products and programmes are in many cases at least, at a certain degree customized or tailored for their primary customers.

References

1. Wahlström, J., Hagberg, M., Toomingas, A., Tornqvist, E.: Perceived Muscular Tension Job Strain, Physical Exposure, and Associations with Neck Pain among VDU Users; a Prospective Cohort Study. Occup. Environ. Med. 61, 523–528 (2004)
2. Hurwitz, E., Carragee, E., van der Velde, G., Carroll, L., et al.: Treatment of Neck Pain: Noninvasive Interventions. Results of the Bone and Joint decade 2000-2010 Task Force on Neck Pain and its Associated Disorders. Eur. Spine J. 17(1), 123–152 (2008)
3. Green, B.: A Literature Review of Neck Pain Associated with Computer use: Public Health Implications. JCCA 52(3), 161–168 (2008)
4. Luck, R.: Dialogue in Participatory Design. Design Studies 24, 523–535 (2003)
5. Rissanen, M.: Customer's Voice in eHealth Evaluation. In: Proceedings of 4th International Future Learning Conference, pp. 230–244. Istanbul University Publication 5115 (2012)
6. Jacobs, C.: The Evaluation of Educational Innovation. Evaluation 6(3), 261–280 (2000)

Creating a Healthcare Research Database
Linking Patient Data across the Continuum of Care

Denise Magnani and Michael Freed

General Dynamics Information Technology,
Health Analytics and Fraud Prevention,
West Des Moines, IA, United States, and London, England
{denise.magnani,michael.freed}@gdit.com

Abstract. Business Intelligence (BI) is a set of methodologies, processes, architectures, and technologies that transform raw data into meaningful and useful information. BI enables more effective decision-making through strategic, tactical and operational insights. Our unique approach and patient linkage allows for improved information delivery, as well as increased access to information, improved ease of use, provision of a platform for collaboration and knowledge-sharing, and the ability to process and present reports from millions of rows of data in a matter of seconds.

Keywords: warehouse, healthcare, quality, research, analytics, linkage.

1 Introduction

Linking multiple disparate healthcare data is critical and invaluable to improving the quality of care and reducing costs and promoting appropriate utilisation. Creating a data-driven view across the continuum of care allows for more indepth monitoring of health trends, public health research, performance monitoring, and policy decision making. We have created a unique patient linkage algorithm and procedures that facilitate accurate and efficient analytic and reporting capabilities. We have created a health data warehouse containing over 255 billion records. This data warehouse includes inpatient and outpatient claims data as well as enrollment, administrative, drug and death certification data for the last 12 years. Through this multidimensional database, we create business intelligence through the analysis of billions of rows of live data from numerous disparate data sources.

2 Framework

Our unique design includes multifaceted beneficiary matching processes, data encryption methodologies, and complex algorithms, which serves as the primary data source for national health research and analysis. The unique star schema design of the warehouse allows for ease of operations, querying capabilities, analysis, integration of new data sources, and leveraging of existing technologies.

G. Huang et al. (Eds.): HIS 2013, LNCS 7798, pp. 277–278, 2013.

3   Business Intelligence

Our BI investigates patterns of cost, quality and service use in populations. The work includes developing and applying statistical methodology for calculating composite quality values using already developed quality measures and data sets, standardising and adding payment data, drug data, claims data, and multiple clinical care setting data.

Author Index

CPSIA information can be obtained at www.ICGtesting.com
Printed in the USA
LVOW07s2017170813

348399LV00005B/134/P